OPPORTUNITIES in Forestry Careers

REVISED EDITION

Christopher M. Wille

VGM Career Books

Chicago New York San Francisco Lisbon London Madrid Mexico City Milan New Delhi San Juan Seoul Singapore Sydney Toronto

The McGraw·Hill Companies

Library of Congress Cataloging-in-Publication Data

Mueller-Wille, Christopher.
Opportunities in forestry careers / Christopher M. Wille.—rev. ed.
p. cm. — (VGM opportunities series)
ISBN 0-07-141151-8
1. Forests and forestry—Vocational guidance. 2. Forests and forestry—Vocational guidance—United States. I. Title. II. Series.

SD387.F6M84 2003
634.9′023—dc21 2003045067

1 2 3 4 5 6 7 8 9 0 LBM/LBM 2 1 0 9 8 7 6 5 4 3

ISBN 0-07-141151-8

Interior design by Rattray Design

This book is printed on acid-free paper.

Contents

Foreword

How would you like to work in a field that manages forests and their ecosystems for the benefit of present and future generations? Working with issues such as climate change, the effects of climate change on forests, how forests can contribute to climate change, the health and growth of trees, and the management of forests, a career in forestry can be very rewarding. Working in forestry can take you outdoors to an urban forest or deep into remote forests; it can take you to the laboratory, into communities, or to the mill. If you enjoy the wonders of science and the beauty of our natural resources, a career in forestry may be of interest to you.

Forests are among the world's most valuable resources. The "2001 State of the World's Forests" report estimates forests cover 3,870 million hectares of the world. Forests provide the oxygen that we need not only to breathe but also to help clean the air. Forests also moderate our climate, regulate our water systems, help prevent erosion, provide habitat for wildlife, and offer recreational and spiritual opportunities. Without forests, landscapes would be dismal and an outdoor artist's canvas would be blank.

Early settlers depended on the forests for their very survival. More than four centuries ago settlers considered the forest a nuisance, and they cleared forested areas for homesteads and farmland. Native peoples used the forest for gathering food and for their spiritual needs. As time passed, forests and the trees within them were identified for commercial needs such as masts for ships, home construction, and paper. Early in the 1900s, forest practitioners started reforesting and planning forest management.

Today, concern about our environment is global. Communication is instantaneous. The World Wide Web has opened its doors wide to information. As the world population increases, there is more demand on the forests and environment and there is a greater need for sustainable management of forests. Recreation and tourism are of great importance to society in North America, whereas in developing countries, the reduction of deforestation and the protection of the forest for basic human needs are of utmost concern. The challenges for those in forestry over the next couple of decades will be greater than before, as we try to meet society's demands as well as protect forests and their ecosystems. It is our generation and the next one that will need to work together and plan for it.

The importance of forests leads to some interesting career possibilities. The forest practitioner works to manage the forests so as to meet our present needs without jeopardizing the needs of future generations. Modern forestry aims at the sustainable management of forests for society as a whole. A forestry career is not just related to the northern wilderness; even large cities contain urban forest settings. Forestry is global and careers can span many continents. Foresters, forestry technicians, technologists, and other forest workers are needed to manage forestlands and to provide a variety of related tasks. Some working in and with the forest use the most advanced technological equipment, such as satellite imagery and computer modeling. Obtaining skills in these areas requires special training and the right combination of interests, attitudes, and abilities.

It is not for everyone, but for the right person forestry can provide a career that is challenging and demanding. A forestry career can be diverse from one day to the next, like the forest itself. If you think you have the potential for working in forestry, this book provides the information you will need to lay the groundwork for such a career. The author has outlined details you will need about types of jobs available, job expectations, background and education needed, and other factors involved in a forestry career. Use this book as an introduction to the career possibilities in this exciting occupational area. Who knows? Perhaps one day you will join the ranks of North American forestry practitioners and work in a very rewarding profession.

Roxanne Comeau, RPF
Executive Director
Canadian Institute of Forestry/Institut forestier du Canada

Preface

Although the total number of foresters in the United States is relatively small (an estimated fifty thousand), their work affects the economic and environmental well-being of the entire nation and the world. *Opportunities in Forestry Careers* details how you can become a part of this important profession.

Forestry began in the United States about a hundred years ago. At that time, many of the forests of the United States had been cut down and burned over, and no care was taken to prevent damage from insects, disease, soil erosion, or fire. The healthy forests of today are due in large part to the development of forestry; in fact, there is now almost as much forestland in the United States as there was when explorers first reached the shores of North America. If you decide to pursue a career in forestry, you will be part of this proud tradition. But that only tells part of the forestry story.

In the last two decades, Americans have come to understand that forests play a significant and critical role in helping to ensure clean air and water, habitat for all forms of wildlife, and a place that each of us can go for spiritual renewal. In today's world of expanding population, there is an increasing need to make certain our forests can support both the biological and economic demands placed on them. That is truly the job of the professional forester, and it places the profession at the forefront of the single most important issue we face in the world today—sustainable development.

It is the mission of the forester to apply sound science to meet the needs of the specific landowner (public or private) and in so doing, help to achieve economic and biological health for the nation and the local community.

After reading this book, you may want to talk to individual foresters about their work. This will give you an idea of their day-to-day responsibilities and the rewards they encounter. Almost everyone will tell you about the importance of interacting with people and keeping up with a rapidly changing world. The appendixes of this book give you some useful ideas for follow-up contact, including the address of the Society of American Foresters.

Members of the Society of American Foresters work for literally hundreds of organizations and companies, large and small, in all fifty states and in many foreign countries. Some spend most of their time out in the woods; some work in labs or offices. Many live in rural areas; others live in New York, Washington, San Francisco, and other cities. Your personal and professional interests can help tailor a forestry career that you will find most satisfying.

I hope what I have written reflects my own love of the profession of forestry and my excitement about what lies ahead. It is a profession where you can truly make a difference in the future of America and have the opportunity to work side-by-side with some very dedicated people.

William H. Banzhaf, C.F., C.A.E.
Past Executive Vice President
Society of American Foresters

1

Overview: Working as a Forester

There are few pleasures in life as great as walking through the forest, breathing in the scent of pine needles, or looking up to see the tops of the trees swaying in the wind. It is no wonder that forestry has always been a profession that appeals to those who treasure such natural beauty. It is a career path that calls to thousands of high school seniors as they contemplate what to study in college. It is also a field that beckons those who find themselves dissatisfied in other professions and longing for work that is meaningful and challenging.

Of course, working in forestry means more than getting paid to hunt, fish, tromp around in the woods, and commune with nature. Many foresters do occasionally have the opportunity for such blissful retreats, but being a professional forester involves more than going for a good hike. It involves using one's intellect to care for, develop, and harvest forests in a way that is beneficial to society and the environment.

Forestry is a career that requires real preparation. You must complete a demanding four-year program at a college or university. Most foresters these days have at least a bachelor's degree, but many have gone on for a master's or doctoral degree. There also are opportunities for forestry technicians who have completed one- or two-year programs at technical colleges or "ranger schools." However, the professional forester, with a degree and license, has a better chance of moving up the ladder to more responsibility and higher pay. Regardless of education, forestry offers plenty of satisfying and interesting jobs for those who are willing to work and willing to learn.

Part of forestry's allure is its romantic history. There is something endlessly appealing about the traditional image we have of the forester. In that image the forester is a healthy, rustic sort who works in the beautiful woods, helps wildlife, and rescues lost hikers. Many of our ideas are based on television and novels. These fictional foresters live in the Great North Woods; they wear plaid shirts and suspenders, carry axes, and consort with brawny loggers. Or perhaps they are solitary figures, holding lonely vigils in fire-lookout towers. And the fictional foresters are invariably men.

Although the modern forester may do all of these things, math skills are probably more important today than the ability to use an ax. Today's forester, whether male or female, may indeed know how to find food in the woods, but he or she must also understand the chemical composition of soils and know how to calculate the strength of bridges.

Some of our ideas about what forestry is and what a forester does may be outdated, as we shall see in the following chapters. But there's no doubt that forestry is sometimes adventurous and that much of the work is done outdoors. To many young people, that sounds much more inviting than a desk job.

Another reason forestry is popular is that it provides a chance to help protect the environment. Since ecology became a household word in the late 1960s, people have become increasingly aware of environmental problems, especially during the last decade. As environmentalism has grown and matured, more and more people want to become actively involved. Foresters could be called professional environmentalists.

Forestry also provides an opportunity to serve humanity. People need wood. We build our homes with it and make paper from it. We use wood to fashion innumerable everyday items, from pencils to pianos. And, of course, wood may even help us solve our energy problems.

As the population of the world continues to increase, we will need more and more wood. Fortunately, wood is a more renewable resource than oil, coal, or iron ore. We can harvest a forest and then help it grow back quickly. Part of the forester's job is ensuring that we have wood for the future.

While our need for wood and its many products is growing, the forest we have is shrinking. We are clearing forestland at an alarming rate for houses, highways, shopping centers, cropland, and other uses. One of the forester's challenges is to grow more wood on less land.

Modern forestry is more than just growing trees. It is an ecological science. Forestry combines timeless natural processes with the latest technology. It is the study and management of a large natural system—the forest—and everything in that system. Foresters have learned to manage a forest system so that it produces lots of wood and serves many other uses at the same time. Forestry includes the administration of the soil, water, wildlife, and people, as well as the trees of a forest.

The demand for forest products is so great that the natural forest can no longer keep up. We have to give nature a helping hand. Foresters plant quick-growing seedlings. The young trees are fertilized, thinned, protected from fire and insects, and harvested according to a production plan. One forested acre, left to nature, might require 250 years to grow a stand of mature trees. That same acre, properly managed, could grow a harvestable stand in fifty years with six times the volume of wood of the untended forest. We are learning how to make a forest produce more wood. At the same time, we are discovering new and improved uses for wood. Scientists are developing fast-growing "supertrees." Other researchers are developing new wood products or ways to use wood more efficiently. Wood, an ancient building material, remains an essential construction medium today. And wood, our first source of heat, continues to contribute to our energy supply.

Forestry offers an opportunity to contribute to the good of society, to serve the natural environment, to work among the splendors of nature, and to have a productive, satisfying career. If that sounds like an appealing combination, forestry may hold an exciting future for you.

You can have a career in forestry in a tremendous variety of jobs. Foresters manage timberlands; conduct research; teach; supervise mill operations; run wood products businesses; serve as advisors, consultants, and product salespeople; act as land planners; work in government bureaucracies; administer parks; and educate the public through the media.

Foresters also work in a variety of places. You will find foresters working deep in the woods, in logging camps, in offices, in laboratories, at computer terminals, in classrooms, in government agency headquarters, in city parks, in Christmas tree farms, in paper mills, in factories, and in city halls.

Government employs many foresters. In the United States, the U.S. Forest Service hires more foresters than any other organization in the country. State, provincial, country, and city governments employ foresters. Most large timber companies also hire forestry school graduates. There are jobs available throughout the wood products industry. Other foresters are self-employed—hiring themselves out as consultants to companies or individual woodlot owners. Some teach in universities, community colleges, or technical centers. Some put their education to good use in a related field such as landscape architecture or horticulture.

The forestry profession includes a large number of specialties. Most people think of foresters as timberland managers doing such things as fighting fires, planting and measuring trees, counting wildlife, and surveying trails. That is one specialty—forest management. Other specialties include engineering, research science, wood products, paper science, wildlife biology, range management, wood chemistry, computer science, forest business, international forestry, urban forestry, outdoor recreation, and communications.

This book explores these forestry jobs, the different places where foresters can work, and various specialties. You probably will be surprised at how many different things being a forester can mean. We will discuss primarily those kinds of jobs a graduate of a forestry school might consider. We will look closely at the many opportunities for professional foresters and at the outlook for employment.

The duties and the outlook for technicians, ranger school graduates, and laborers will be covered. Since some of the better forestry career opportunities are with wood utilization operations, jobs in pulp and paper manufacturing and the wood products industry also will be surveyed.

We also will discuss what college will be like and how it differs with the specialty you select. This will help you decide how to focus your education. Appendixes C and D comprise a complete list of forestry and technician schools. After you decide what you want to study, and where, you can write to several schools for detailed information. Important forestry associations and periodicals also are listed in the appendixes.

Forestry is a combination career. It combines outdoor work with office duties, dealings with people and with nature, production with protection. Timeless natural cycles and ancient techniques are combined with the latest technology and research. These combinations keep forestry jobs interesting and diverse. The various jobs require different kinds of people with different skills. Forestry is also a challenging and demanding profession. Meeting these challenges can be very satisfying. If you are interested in the earth's future as well as your own, consider one of the many rewarding opportunities in forestry.

2

HISTORY OF FORESTRY

THE ROOTS OF forestry are deeply intertwined with the history of mankind. From the earliest days, our ancestors relied on the nuts, fruits, roots, and leaves of trees for their survival. Shelters were constructed from wood, and wood was used to build cooking fires. The things that make us human come from our ability to fashion wood. As far as cultural achievements, the first writing was done with a burnt stick and the earliest known wheel was made of wood. And when our predecessors sought food or fought with one another, they fashioned many of their early hunting tools and weapons from wood.

Although our evolution and survival has always been closely linked to our use of wood, forestry as a profession is a more recent development. How can this be? Remember that our definition of forestry is the scientific management of forests as a renewable resource. Unfortunately, this wise-use management of our forests has not been practiced until relatively recently in human history.

As early people gave up wandering and hunting in order to farm, they needed land crops. The best land had trees, so people learned to burn forests to expose the rich soil. When the farmed land was exhausted, the people would abandon it and clear another forest.

People began living in villages, which soon became crowded. Wood became increasingly valuable as settlements grew. Some historians think that China suffered shortages of wood as early as five thousand years ago. The Greek and Roman Empires also needed more wood than they had. By the Middle Ages, some countries in Europe had banned the clearing of certain forests. The royal forest was always protected so that the king would have good hunting grounds.

John Evelyn's *Silva*, written in 1662, is considered by many to be the first forestry book ever published. At the time, wars in England had led to the depletion of the old forests, and the shortage of good wood was threatening the success of the monarchy, which had just been restored. Trees were needed for the nation's defense. For example, building a seventy-four-gun ship could take up to thirty-eight hundred trees, or about seventy-five acres of timber. But these trees took about one hundred years to grow. Faced with this crisis, the commissioners of the navy turned to the Royal Society for guidance on managing woodland. John Evelyn was commissioned to produce a study about how trees could be cultivated quickly. He created something akin to a handbook, which describes the purposes and practices of timber management and gives details on the different species of trees, such as how to grow them and how trees should be pruned, harvested, and seasoned for use.

During the nineteenth century, forest management developed rapidly in Germany, France, and Scandinavia. This was really the beginning of the science of forestry. The Europeans devised plans to control harvesting, to prevent soil erosion, and to maximize timber production. By 1825 there were formal forestry schools in Europe. Across the Atlantic, in newborn America, a different history of forestry was unfolding.

Forestry in the United States

The early American colonists found endless unbroken stands of leafy green hardwoods covering the eastern seaboard. The forest stretched its lush blanket all the way to the Midwestern prairies. There trees gave way to dense blue and amber grasses that grew so tall that a person on horseback would sometimes have to stand in the saddle to see over them. Huge conifers shaded the Rocky Mountains and the West Coast. In all, nearly one-half of the North American continent was forested.

America's forests were an appealing treasure to the European settlers. The year after the *Mayflower* arrived, a ship was sent back to England loaded with timber. Soon, legions of ships were taking American logs and lumber back across the Atlantic.

With the aid of fire, the settlers eventually cleared millions of acres. As the frontier villages grew into towns, more and more wood was needed for fuel and construction. Some wise colonists, like Benjamin Franklin and William Penn, became concerned about the rapid disappearance of the forest. Concern over shortages of firewood encouraged Franklin to build one of his many inventions, the Franklin stove, which burned wood economically. William Penn made what some consider the first official attempt at forest conservation in this country. He wrote a clause in land contracts requiring that one acre of trees be left for every five acres cleared.

At the time that the United States gained independence from Britain, tremendous amounts of wood were being used to build farmsteads, mines, railroads, industries, and cities. Sawmills sprang up everywhere to fill the growing need for boards and timbers. Wood became scarce in some places. The colonies were soon arguing over who owned the forests. To prevent these arguments, the federal government began buying forestlands. These lands, being in the public domain, belonged to everyone and were to be used for national purposes. Before long, the federal government owned three-fourths of the continental United States and all of Alaska—1.8 billion acres.

The government had more land than it knew what to do with, so it began giving land away to homesteaders, railroads, timber companies, mining companies, and others. This caused the greatest land rush in history. Millions of acres were given away, often to the person with the fastest horse, the quickest gun, or the loudest lawyer.

The federal government of the United States still holds much beautiful and valuable property. These lands are managed by various government agencies. The Bureau of Land Management, for example, manages about 270 million acres. The Department of Agriculture, which includes the Forest Service, administers 191 million acres. The Army Corps of Engineers manages more than five million acres. These federal bureaus employ more professional foresters than anyone else.

Even though vast tracts of forest were set aside as a kind of national resource bank, the waste of America's forest continued. There were no laws protecting the public lands. There were no professional foresters to plan how they should be used. Forests were still being burned. Logging practices wasted as much timber as was harvested. There were no efforts to reseed the cut or burned areas. Valuable tree-growing soil was left open to erosion.

Conservation

It was not until the middle of the nineteenth century that forestry management as we know it today began to be practiced. In 1847 George Perkins Marsh, a congressman from Vermont, published an address that advocated a conservation approach to the wilderness. In the 1850s and 1860s, the speeches and publications of Henry David Thoreau advanced the conservation ethic. Finally, during the last quarter of the 1800s, concern for our forest resources began taking root. In 1872 the governor of Nebraska proclaimed the first Arbor Day. Dedicated to planting useful trees, Arbor Day is now celebrated in the spring in every state.

The next year, a federal law was passed that gave free 160-acre prairie homesteads to anyone who would plant trees on 40 of the acres.

In 1876, a full century after the Revolution, Congress commissioned a scientist to survey the American forest. For the first time, lawmakers had some idea of the true value and extent of our forests. Some states responded by forming forestry boards or commissions. In 1881 Congress established the first Division of Forestry in the Department of Agriculture.

Still the public lands had no protection. An important law was passed in 1891 that gave the president authority to set aside forest preserves. Within six years, forty million acres had been designated. The Department of the Interior had the power to sell timber from these lands and to protect them. But there were still no foresters and no scientific management of public lands. Then some universities began offering courses in forestry modeled after European programs. Yale opened the first school of forestry in 1898.

Natural resource management got a real boost in 1901 when Theodore Roosevelt became president. Roosevelt was an expert hunter and avid outdoorsman. As president he began the wildlife refuge system and added greatly to the forest system.

Roosevelt was influenced by his friend Gifford Pinchot. Pinchot was head of the Division of Forestry when that agency became the Forest Service in 1905. He was a farsighted and energetic administrator. Through the efforts of Roosevelt and Pinchot, the American public was introduced to the concept of "wise use" of natural resources.

The Forest Service remained part of the Department of Agriculture. Management of the forest reserves then was transferred from the Interior Department to the Department of Agriculture. These reserves, totaling one hundred million acres at the time, were named national forests. Most of the national forests were in the west. In 1911 Congress authorized purchase of additional public forests in the east for timber and watershed protection.

World War I interrupted the progress of American forestry. There were still only a handful of professional foresters in the country. Some of them spent the war years in Europe helping produce the wood needed by the fighting troops.

After the war, forestry, like everything else, struggled through the Great Depression. Another Roosevelt, President Franklin D. Roosevelt, began ambitious public works programs. One of these programs, the Civilian Conservation Corps, employed hundreds of young people in forestry projects. They planted trees; fought fires;

built roads, bridges, and lookout towers; and made other improvements on the public forestlands.

World War II brought these programs to a halt. Foresters again turned their attention to producing the wood products needed to fight a war.

After the war, the U.S. Forest Service quickly grew into a large and progressive organization. One factor in this growth was that the American economy entered a long period of prosperity. New homes were built at a record rate. More wood and wood products were needed than ever before. The science of forestry blossomed. Research was conducted to find the best forest management techniques and to discover new ways to use wood.

The federal government was not the only organization developing the science of forestry. Almost every state had a forestry program of some kind. States purchased and managed forests for the benefit of their people. By 1949 state forest holdings totaled 14.8 million acres.

Private industry also was attracted to wise-use management of forests. This concept holds that forests, like any other crop, can be harvested and renewed each year, year after year. With proper management, a forest can continue to produce a high volume of wood virtually forever. The concept of sustained yield was now widely known. It was clear that demand for wood would easily outpace what the company-owned forest would produce naturally. New methods of harvesting, replanting, and woodland protection were developed. Timber companies were hiring foresters to ensure the future of their business.

Both government agencies and private timber companies were investigating new frontiers in forestry. More and more foresters with different specialties were needed. Colleges across the country opened forestry schools. Forestry was always a popular profession, but since the dawning of the environmental era of the late 1960s, it has become even more popular.

The attraction to forestry as a career is certainly understandable. It offers tradition; it is interesting, sometimes exciting work; it provides a chance to work in the clean, healthy outdoors; and it even has a touch of glamour. Best of all, forestry is a young and dynamic field where a person can make his or her own opportunities. The real history of forestry is being written right now.

Forestry in Canada

Forestry is a major industry in Canada. More than nine hundred thousand Canadians are employed, either directly or indirectly, in the forest products industry. Half of the country is forested, and Canada contains 10 percent of the world's forestland, from which flows 20 percent of the world's fresh water.

Throughout its history, Canada's economic growth has been tied to the success of the forestry industry. In addition to newsprint, Canada manufactures other papers, cardboard, and other wood products such as furniture. Wise management and conservation of forestland is important to Canada's future.

The goal of Canada's foresters is to manage this natural resource so that it yields a combination of economic and environmental benefits. The need to harvest wood is balanced against the need to protect wildlife and the desire to provide a place for recreational activities such as hunting, fishing, and camping.

The Canadian public owns 90 percent of the country's forest. The major employers of foresters in Canada are federal, provincial, and municipal governments, as well as private industry. Interesting opportunities are available abroad, especially in the Third World, and these can be sought through such organizations as the Canadian International Development Agency (CIDA) and the International Development Research Centre (IDRC).

Competition for forestry jobs is increasing in Canada. However, increasing public awareness of the need to manage forestland wisely and the demand for greater efficiency in forest resource management will ensure a continuing need for competent foresters. For more information on forestry in Canada, you can contact the Canadian Institute of Forestry, 151 Slater Street, Suite 606, Ottawa, Ontario, Canada K1P 5H3.

3

The North American Forest

The definition of a forest is fairly straightforward. It's an area covered with trees. In general, what scientists refer to as *forestland* is an area that when left alone for many years will naturally grow trees.

Exactly what sort of trees and plants flourish in a given area depends on a number of factors, including rainfall, temperature, soil, geographical location, and terrain. Left to its own without human interference, an area will grow the plants nature intended. A distinct area—such as a forest with all its plants, animals, and other environmental elements—is called an ecosystem. Nonforest ecosystems, then, would include prairie, tundra, beach, desert, and cropland.

That is not to say that we cannot grow the plants from one ecosystem in another. For example, trees native to one particular region have been transplanted to another, and forests have been cut down to grow agricultural crops. Nevertheless, it would be very difficult, if not impossible, to force a forest to grow on the prairie or the desert.

In Canada about half of the land is covered by forests, and forestry is a major part of its economy. The United States contains more than seven hundred million acres that are considered forest. That is an area nearly equal in size to the entire United States east of the Mississippi River. More than five hundred million of these tree-covered acres are classified as "productive" forest. These vast areas are capable of growing a profitable amount of wood. The most productive can produce more than twenty cubic feet per acre a year.

The Nature of Trees

Trees are the most noticeable and familiar plants on earth. In addition to making up forests, they add shade and beauty to our yards, soften the harsh lines of city streets, and add variety to the rolling fields of the countryside. Most wildlife species depend upon trees for food or cover. And, of course, they and other green plants release the oxygen we need to survive.

Botanists define a tree as a woody plant with a single, erect stem growing to a height of ten feet or more. It has a perennial trunk, which means that it continues growing year after year, rather than dying back each season like many garden plants. Shrubs are woody plants, too, but they are generally shorter and have branching stems rather than a single trunk.

Trees can grow almost anywhere. There are over fifty thousand tree species worldwide, and more than nine hundred species of trees growing wild in the United States. About 160 species are considered commercially valuable. Florida alone has 314 species of trees, more variety than any other state. Texas, Georgia, and California also have amazing numbers of tree species.

Some trees are small and wispy when full grown. Others are giants. In the lush, wet forests of the Northwest, Douglas firs often grow to more than two hundred feet.

Both the oldest and largest living things on earth are trees. The General Sherman tree, a giant sequoia in California, is more than 270 feet high, 115 feet around, and more than three thousand years old. The tallest redwood ever recorded was 364 feet, which is almost as tall as a forty-story building. A Morton Bay fig tree growing in Santa Barbara, California, is almost seventy feet tall and its trunk is more than thirty feet around. The tree was planted in 1877 from a seed brought over from Australia. It has such spreading boughs that about fifteen thousand people could stand in its shade! Perhaps the oldest living thing on earth is a bristlecone pine growing in the White Mountains of California. Foresters have estimated that this living relic is forty-eight hundred years old.

Some of the largest trees contain enough wood to build dozens of homes. Yet these huge plants grow from seeds so small that several hundred will fit on a teaspoon. Trees are miraculous factories that can turn sunlight, water, and minerals from the soil into a great column of wood and millions of leaves. Trees produce huge volumes of oxygen and consume carbon dioxide. They also have a kind of cooling system. Trees suck hundreds of gallons of water through their roots, pump it up the tall trunk, and release it as vapor through pores in their leaves.

Some botanists say that we see only half a tree. The other half is underground in the spreading network of roots. An oak thirty feet high may have roots growing out sixty feet from the base of the tree. Roots anchor a tree to the soil. That is important to the tree, of course, but it is important to us, too. All those miles of roots hold the soil in place. Without plant roots, soil is quickly washed or blown away by the elements.

Roots also absorb minerals and water and transport them to the tree. Some roots store food until needed. Roots continue to grow throughout the life of the tree.

Different species of trees have different kinds of wood. Some woods are soft, some hard. Some will bend into almost any shape, while others are brittle and snap easily. Some are very light. Others, like teak, are so heavy they sink in water. Foresters learn to identify woods by their look and feel. Many woods can be recognized by their smell. Since each kind of wood has special value and uses, it is important for the forester to know each of them.

Most people recognize trees by their leaves, fruit, or flowers. Foresters know these characteristics and can identify a tree by its bark, by its shape, or even by where it is growing.

A forest is made of lots of individual trees that live together as a group—a community. Forests are like cities where everything depends on everything else. And, like cities, each forest is different. Foresters learn to recognize forest communities, or ecosystems. Each is valuable in a different way, and each must be managed in its own special way. The forest manager must understand how trees, wildlife, soil, and climate affect one another.

The Many Uses of Forests

Ask ten people what the most important part of a forest is, and you probably will get ten different answers. The logger will say wood. A hunter might think that wildlife is the best thing a forest produces. A city dweller may prize the peace and quiet. An artist might answer beauty. The forester will give all of these answers and many more.

Foresters understand that the forest has different meanings to everyone. They know that it looks different through the eyes of a hiker or photographer than it does to a real estate developer, skier, or bird-watcher.

One of the most important responsibilities of a forester is to seek a way to balance the different expectations that people have. It is a challenging balancing act. The forester must figure out how to get timber and boards out of the woods without ruining the other assets or values. It is possible to harvest logs out of an area without destroying the wildlife, wrecking a stream, or spoiling the scenery forever.

As a student in forestry school, one of the first things you would learn is the "multiple use" concept. This is the management system that tries to get the greatest good from our forests for the largest number of people. It requires accommodating the various uses in a way that is satisfactory to the various needs of people and wildlife.

Under skillful management a forest can produce saw logs, be a home to wild animals, and also be a popular hiking area. The same woodland can serve as a watershed, a seasonal grazing range for livestock, and a place to hunt, camp, fish, or cross-country ski. We have to make the best multiple use of our forests because our growing population makes ever-greater demands on them.

Exploitation, Multiple Use, and Wilderness

Trees are different from other resources in one very important way: they are a renewable resource. It takes millions of years for deposits of oil, copper, coal, or iron ore to form. Because of the length of time involved they are not renewable resources. If properly managed, however, living resources like trees, grass, and wildlife will continue to produce for a long time. We can harvest trees or game animals as we would a crop of corn or wheat, because they will grow back.

Foresters decide how many trees to cut in an area, when to cut them, and how to build the transportation system to get the logs out of the woods. They supervise replanting, fertilization, fire control, and animal damage control. Foresters make sure that streams and lakes are not damaged by logging operations and that the soil is protected from erosion by wind and rain.

Foresters also determine what other activities can occur at the same time in the same woodland. While the area is being logged, the public might be allowed to salvage firewood. As the new young trees are growing back, the forester may open the area to hunting or limited cutting for Christmas trees. This is multiple use at its best.

Decades ago, loggers would simply go in and clear-cut an area, taking out every tree in sight. They gave no thought to soil or stream protection. They did not bother to replant the site or leave special places for wildlife. This was uncontrolled exploitation. Some loggers still practice this shortsighted "cut and run" method, but it is now rare. By now we've learned that our demands for wood cannot be met unless we manage our forests for sustained yield. And much of our forestland, both publicly and privately owned, is now administered by foresters to produce the maximum amount of wood for today's needs as well as for the future.

At the other end of the management scale, we have wilderness. There are a few places left in the country that remain unchanged by human activities. Some of these precious areas are set aside as wilderness. Although some commercial activities are allowed in the wilderness, every effort is made to leave these areas untouched.

Wilderness is special to foresters. Only in these natural areas can foresters see nature take its course. The wilderness is a natural laboratory. It is special to all of us for different reasons. We treasure the opportunity to walk in countryside that has never felt the ax, plow, or bulldozer. It warms the heart and lifts the spirit to experience solitude in nature's untouched majesty.

Few industrialized countries have wilderness. It is a luxury prized like a diamond for its rarity. Only a small amount of land in the United States is protected as wilderness. This protection generally prohibits logging. Wilderness is, nonetheless, a multiple-use area. Wilderness is an excellent watershed protector and wildlife habitat. Hiking, camping, and scientific study are other uses of wilderness.

The Forester as Scientist and Politician

In the middle of the last century the environmental movement truly caught the public's imagination. Since 1900 we have realized that there are no more frontiers on earth, no more lands to discover. But in the 1960s this awareness blossomed into a more generally recognized imperative to make the best use of nature's gifts. This environmental movement continues today.

Most people would not condone the uncontrolled exploitation of natural resources anymore. Yet even though it would be rare to find someone who advocated the outright destruction of our environment, there is still a lot of room for disagreement over the meanings of conservation and multiple use. Well-informed and well-meaning individuals still can bitterly disagree over how much timber should be harvested and by which method. Arguments rage over use or preservation of wilderness. Some say we need more preservation; others claim we could get by with less.

As a professional forester you will be functioning as a scientist. However, you will not be able to avoid the politics of resource management. As the competition for our resources increases, foresters will often find themselves in the center of public debates. Today's foresters are part scientist, part politician. At times they also must be educators, diplomats, advocates, law enforcers, and trendsetters.

Marketing Forests

The forest is one of the greatest assets of the United States and Canada. Some have compared this forested wealth to Saudi Arabia's oil riches. There is one big difference: oil wells eventually go dry and the resource runs out for the foreseeable future. Forests, if properly managed, can continue to produce wood products indefinitely.

Foresters sometimes divide the American forest into four regions for purposes of discussion. The regions have different tree species, produce different wood products, and require slightly different management techniques. These four regions are the Northeast, the Southeast, the Rockies, and the West Coast.

The forests in the Northeast were the first to be harvested. This is where logging got its start in America. The Great North Woods are composed mostly of hardwoods: oak, maple, birch, beech, and aspen. Almost half of the nation's commercial forestland is in the Southeast. Much of the country's pulp and paper wood comes from the southern "wood basket." High-quality pines such as loblolly, slash, and longleaf grow to harvest size in twenty-five years with proper management. Half the southern forest is composed of hardwoods such as oak and hickory. The Rockies, from Canada to Mexico, comprise a broad band of conifers. Ponderosa pine, Englemann spruce, white fir, and larch are common species. The West Coast is the land of the tall timber. These awesome stands of softwoods include fir, hemlock, cedar, spruce, redwood, and pine trees. Most of the western trees are evergreen conifers. The columnar giants require fifty to one hundred years to mature.

There are more than five hundred million acres of potentially productive forest in the United States. These acres are either producing wood now or could be if their owners chose to put them into wood production. Who owns all this tree-covered land? Private individuals own more than half, about 58 percent. Through the federal, state, and local governments, the public owns 29 percent: 22 percent of this is federal lands, while 7 percent is under state and county jurisdiction. The forest products industry owns 13 percent. Although the timber industry has the smallest share, its lands produce almost 30 percent of the forest products used. The numbers emphasize the effectiveness of professional forest management. Because industry is managing its land primarily for wood production, its acres are much more productive than are the private or government acres.

Roughly one acre of American forest is in commercial production for every person in the country. That sounds like it might be enough, but two factors are nibbling away at each acre. First, there is population: the American population continues to grow, now topping 287 million, and it is expected to rise to 346 million by 2025. Second, there is land conversion: every year more than one million acres of commercial forestland are cleared and made into something else—cropland, grazing range, housing projects, or other developments. That, too, makes each piece smaller.

However, American commercial forests could be producing twice as much wood without damaging the environment. How?

With careful, professional forest management. The forester learns how to apply two key principles. Those principles are productivity and efficiency. Foresters are trying to share those secrets with every individual, company, and government organization that owns timberland.

Making the Most of Forested Land

Although we cannot create additional land, we can better utilize the land that we have. American farmers are famous for figuring out how to get maximum production from every acre of land. Foresters can do the same thing and even use many of the same techniques. At the moment, however, many foresters would agree that we're not using our timberlands in the most effective and creative ways.

After an area has been logged, the forest manager may decide to help nature replant it. Crews are called in to cultivate the land. Special new strains of trees are planted by machine or by hand. The seedlings are given a head start in a nursery. When planted in the forest, they may already be two or three years old.

As the seedlings greet their first seasons in the wild, the forester may have workers control or cut back other competing vegetation. This gives the young trees every advantage. If soil conditions warrant, the forester may decide to fertilize the trees, just as a farmer fertilizes crops. Later, the forester may thin the stand to prevent crowding. The young trees, like the farmer's crops, are protected from animals and insects.

Although there are similarities between tree farming and crop farming, there are significant differences, too. One important difference is that the farmer can harvest a crop after only a few months. Foresters must wait decades to see the results of their efforts, and thus they must manage today what their children's generation will harvest. The long time frame is one of the aspects that discourages many landowners from trying to grow commercial crops of trees. The private woodlot owner who cuts trees is likely to replant the land with an annual crop, like cotton or soybeans. Why plant trees and wait twenty to thirty years before harvesting again? Foresters and the government must devise incentives, such as tax breaks, to encourage woodland owners to cut and then replant trees.

There are ways to manage a woodlot so that some trees can be harvested every year. That way the lot owner receives a steady income just as if the crop were an annual one like soybeans or corn. Of course, the woodlot has other uses while the trees are growing. It can benefit the landowner by controlling erosion, blocking the wind, protecting a watershed, and harboring wildlife.

The Increasing Demand for Forestry Products

Most people take forest products for granted, but the forester is always aware of two sobering facts. First, the human population is growing rapidly. Second, the amount of commercial forestland is constantly shrinking. The real challenge for the modern forester is to produce more wood products from the same number of trees.

We have done surprisingly well so far. The size of the annual timber harvest in the United States today is about the same as it was back in 1900. That is amazing when you realize that the per capita consumption of paper and production of lumber and other wood products have greatly increased.

Improved Resource Utilization

How have we been able to increase our consumption of forest products without cutting more trees? A forester might answer that question by saying, "improved resource utilization." That simply means doing much more with what we have, which is another definition of conservation.

Through advances in technology, we are using more of each tree. From the time a tree is cut down in the woods until the wood product is in the hands of the consumer, there are almost limitless opportunities to improve efficiency. Loggers are much thriftier these days. Very little of a tree is left behind in the woods. Some operators even move wood-shredding machines into an area being logged. Pieces too small for lumber are shredded and hauled away as chips to make paper or composite board, or they are mulched and laid back down on the trails to provide a nice base for recreational pursuits such as horseback riding, hiking, and snowmobiling.

Sawmills used to waste tremendous amounts of wood. In fact, they burned so much "trash" wood that the smoke caused air pollution. No more. Now all those odd pieces and the bark are used to make dozens of different products. There is almost no waste.

Another key to using wood more efficiently is recycling. Much of the paper manufactured today is made of recycled paper. Instead of throwing the daily newspapers out with the garbage, many families are saving them to be picked up at curbside or deposited at a community recycling center. More and more businesses, schools, and offices are sending their used paper to a recycler to be processed into new paper products. Higher-quality papers are being produced with greater quantities of waste paper. Recycling reduces the number of trees that must be harvested and helps hold paper prices down.

Trees of the Future

Research scientists are inventing still other ways for us to make better use of our forests. Specialists in genetics are studying some of the most exciting possibilities. Scientists are almost daily discovering more of the secrets of how plants and animals pass their characteristics on to offspring. In fact, scientists can now control parts of this process and breed plants or animals with just the desired traits.

All living things carry genes, which are microscopic records of their characteristics. Scientists have learned to interpret the coded information on genes and to combine the genes of different plants and animals. This promising science is sometimes called *bioengineering* or *biotechnology*.

By tinkering with a tree's genes, scientists can "engineer" its offspring. In time, the bioengineers may be able to design supertrees. These trees will have all the characteristics a wood producer could want. They will grow straight and tall in poor soil, they will grow at super speed, and the wood will be of top quality. Supertrees will be resistant to disease and insects and do everything except cut themselves down and climb aboard the log trucks. Bioengineers have already made encouraging progress. Many of the seedlings grown in modern nurseries have built-in advantages over wild trees.

The Green Revolution

While the bioengineers are developing bigger and better trees, other scientists are working on ways to make better use of the trees we have. At one time, only the long, straight, center portion of a tree was useful. This was cut into boards and timbers, and the rest of the tree was discarded. Then we learned how to peel thin strips from a log to make plywood. Later we discovered ways to glue wood chips and sawdust together to make composite boards. Paper mills, of course, also learned to use bits of wood that might otherwise have been thrown away.

Recent improvements in adhesives and in manufacturing techniques have made new wood products possible. Construction materials now can be fabricated from small, low-quality trees that once would have been ignored by loggers. More and more composite wood products are being used in construction. This means less demand for prime logs and greater use of scrub growth.

The advances that have been made in forest utilization are truly remarkable. Yet each new discovery highlights how much more can be done. We will need every innovation if we hope to meet tomorrow's demand for wood products. Forestry still holds plenty of challenges and diverse opportunities for the ambitious and resourceful individual who enters this field.

4

International Forestry

Our world is now characterized by increasing globalization—the exchange of products and ideas across national boundaries and over vast expanses of water. The professional forester has not been left behind while the world has changed. If anything, the expertise of professional foresters is now in even greater demand, required by governments and peoples the world over. One reason, unfortunately, is that development in many countries comes at the expense of their environmental legacy. At this time, forests in many countries are being destroyed at an alarming rate. Some estimate that we are losing more than three thousand square miles of world forest every year. At the same time, world population has grown steadily and now exceeds six billion people. All these newcomers to the planet need wood products, too.

In many nations, the demand for wood is so great that forests are being harvested but not replanted. It's easy to see that this is a dead-end route, akin to spending all of the money you earn as well as your savings. By thus depleting their forest resources, with no provisions for the future, many countries are, in effect, going ecologically bankrupt.

The world's forests do more than provide wood. For example, trees soften the impact of falling rain and hold the soil against erosion. Countries such as India, Nepal, Pakistan, and Thailand have stripped the forests from high mountain areas. Now when it rains, lowland areas are hit by devastating floods.

Forests are home to wildlife. Thousands of species of plants and animals in the Amazon River basin will be wiped out even before they are identified because the lush tropical forest is being cleared to make room for cattle, crops, and settlements.

Forests are an important link in the web of life. They cannot be removed without serious environmental consequences. Yet the pressures on world forests are tremendous. Many of the world's people still use wood as their primary fuel. In some places, wood is scarce and it takes much of the day to gather enough to build the dinner fire. In Africa and in Central and South America, forests are being cleared for grazing land. The developed countries of Europe, North America, and Japan also place demands on the forests of Third World countries. The wealthier countries offer a ready market for beef and wood exports. This encourages the resource-poor nations to strip their forests even more.

The Move Toward Reforestation

It is not as if many governments do not see that their forests are under threat. Indeed, many governments around the world have launched ambitious reforestation programs. Both China and South Korea have reforested thousands of acres. Some African nations have replanted trees to try to stop the spread of the Sahara Desert.

International aid agencies have given reforestation high priority. As an indication of how serious the problem is, the World Bank has provided funds for tree planting in needy countries. The United Nations Food and Agricultural Organization, the U.S. Agency for International Development, and the Peace Corps also are actively involved in reforestation.

Some of the most effective programs have been those that have earned the acceptance and encouraged the cooperation of the local people. Community woodlots, funded by an aid organization but planted and maintained by villagers, have been successful. These woodlots provide the local people with the firewood and construction materials they so desperately need. Sometimes the wood surplus can be sold, bringing welcome cash into the local economy.

Other successful projects have integrated agriculture with forestry. It is possible to grow some trees and crops together on the same land. This technique, called *agroforestry*, produces both wood and food. Some erosion-control projects also have been successful, and there are optimistic reports on experiments with fast-growing tropical trees. Some of the species being tried in Third World countries can grow ten times faster than a Georgia pine.

The International Forester

International forestry requires a forester to be extremely adept at working with a wide variety of people. In most projects, international foresters' work will bring them into contact with a greater number and variety of people than just about any other forestry job. To many people in other countries, reforestation and forest management are foreign ideas brought by foreign scientists. At times, the local individuals need to be convinced of the benefits of a project to win their cooperation and participation. Language and cultural barriers must be overcome. Politics and diplomacy are important. The forester in a foreign country must be adaptable and understanding. The special challenges of working in an unfamiliar environment and culture with new people require an extra measure of ingenuity, energy, and common sense.

Today the role of the international forester is critical. Solutions are needed to slow the quickening pace of world deforestation.

Most North American foresters working overseas are with one of the foreign aid organizations. Some of the larger wood products companies also are doing business abroad. In recent years, there have been a number of exchange programs in which foresters from different countries visit each other's home operations. As the world wood situation grows more critical, we will find ways to export more American forestry expertise and to learn more from knowledgeable tree-growers in other countries.

5

Training and Job Duties

From the woodlot to the boardroom, the sawmill to the classroom—few careers offer as great a variety of opportunities and locations as do those in forestry. The work of forestry also appeals to a very divergent group of individuals. Indeed, it would be difficult to point out a forester in a crowd, since there is no set uniform. The forester could be the person in the plaid shirt and hard hat, the one in the laboratory smock, or even the person in the elegant business suit carrying a briefcase.

And where would one be certain to find a forester? Not all foresters work in the woods. Many spend most of their time behind a desk. Others conduct their business in nurseries, sawmills, manufacturing plants, grasslands, classrooms, and legislative halls.

Forestry is an attractive career to so many people because it contains so many disciplines, allowing one to focus on the area that is most appealing. Future foresters often begin selecting their individual disciplines before the first day of college. That is one of the reasons for this book. It will give you some insight into the various forestry-related occupations so that you can begin planning now. Do not get too concerned about selecting a specialty yet. Most schools will not require that you choose your emphasis until your third year. This discussion will give you an idea of the variety of jobs available in forestry.

Most graduates of a forestry school receive the same degree—a bachelor of science, or B.S. Yet, because there is a real need for specialists, forestry schools give you the opportunity to focus on a particular discipline of forestry. In addition to the basic forestry curriculum, you can select from a broad menu of courses that allow you to study specialized subjects in depth. This flexibility lets you, as an aspiring forester, custom-tailor your education. With the advice of an experienced counselor, you can mold your education to match the projected job market.

Universities often encourage or even require first-year students to stay with a general course of studies. This gives entering students an opportunity to learn about the various forestry professions and sample some of the course work. The well-rounded second- or third-year student has a much better basis for choosing a specialty than does the uninformed incoming student.

When the time comes for you to select elective courses and define your specialty, professors, school counselors, and older students will be around to help you make the decision. No matter which specialty you choose to study, most forestry schools try to make certain that each graduate has a good working knowledge of all the forest science disciplines.

Most forestry school graduates agree that forestry school provides a wide-ranging, useful, comprehensive education. The graduate is prepared to move into one of the many available disciplines. This allows lots of career flexibility. In addition, the graduate is well suited for a successful career in business, agriculture, horticulture, landscape architecture, and the management or marketing of other natural resources.

The degree-bearing forester may also find nonforestry jobs in government, sales, teaching, or research. Given the tremendous variety of fine opportunities within the forestry profession, however, there is usually no need to go outside the field.

In this chapter we will review some of the specialties offered by forestry schools. It's an opportunity for you to discover the type of work each specialist does, the required schooling, and the prospects for employment.

Forest Management

Perhaps when you imagined being a forester you had a particular image in your mind. You might have pictured yourself striding through the sun-dappled woods, dressed in a wool shirt, sturdy boots, and flat-brimmed ranger's hat. Or maybe you saw yourself working in a national park, communicating with headquarters on a portable radio while you monitored the conditions in one particular stretch of forest. No matter what your early image of the profession may have been, you now know that there is much more to forestry than that.

Yet there is one forestry specialization that most closely resembles the traditional image: forest management. This is perhaps the most popular program at every forestry school. It is the curriculum that teaches you a little about everything having to do with taking care of forests. It is the course for those who definitely want to work in the forest. It provides the broad academic base necessary for the general-purpose forester, the jack-of-all-trades.

The forest manager's job is to administer forest resources for the sustained production of goods and services. The forester must consider all the values of a woodlot: timber, recreation, water, wildlife, forage, and scenic quality. The manager's goal, simply put, is to facilitate multiple use.

With a degree in forest management in hand, you might first seek employment with either a government agency or a private timber company. There you will begin as a junior forester with limited responsibility. Then, as you gain experience from veteran foresters, you will be given more and more responsibility. Soon you will be managing your own forest.

Job Duties

The job of a forest manager is often very hands-on. Tasks might include protecting the forest from fire, insects, and disease. Other responsibilities might be planning and managing recreational facilities, developing and protecting a wildlife habitat, and maintaining scenic views. If the area you are charged with managing is being harvested, you will plan and supervise timber sales, arrange land exchanges, and help prepare environmental impact statements. During logging operations, you will be on-site to ensure that maximum production is obtained with minimum environmental damage.

Not surprisingly, this job requires knowing about more than trees. It's an administrative position, so you will supervise technicians and work crews. It's also a public position, bringing you into contact with the general population as well as with politicians and the press. You may be faced with a marketing problem on one day and a political puzzle on the next. It will be necessary to work with specialists such as engineers, geneticists, lawyers, and biologists. As a manager, you must be able to wear many different hats and speak the specialized language of experts. To do all this, you will need a multifaceted academic background. And that is just what you will get in the forest management curriculum.

Education

Working as a forest manager calls for a strong general forestry background, with an emphasis on science and technical training. Most forest management curricula include the biological, physical, and social sciences as well as the professional courses you'll need as a manager. Here is an example of a schedule of classes for the forest management student:

First-year courses
General botany
Calculus
Chemistry
Introduction to computing
English composition
Introduction to forestry
Dendrology (study of trees)
Physical education

Second-year courses
General physics
Principles of microeconomics
Plant physiology
Computer programming/introduction to computers
Soils
Aerial photography

Forest engineering
Mensuration (science of measurement)
Wood technology
Applied statistics
Physical education

Third-year courses
Forest pathology
Forest entomology
Forest ecology
Forest biometrics
Forest recreation decision-making
Logging methods
Forest fire management
Range resources
Group dynamics
Government policy
Forest economics

Fourth-year courses
Watershed management
Silviculture (development and care of forests)
Forest regulation
Forest resource analysis
Multiple-use decisions
Human relations

In general, students lay the foundation of their education with the broad range of basics and then concentrate on a particular discipline. Most schools allow you to register for a fairly large number of elective hours so that you can begin to zero in on the specialty that most appeals to you. Areas for intensive study include forest harvesting, forest products, forest biology, recreation management, range management, wildlife management, business administration, public administration, international forestry, communications, and statistics.

Some schools offer these options as part of the forest management program. If, for example, you specialize in engineering or range management, you still can stay within the basic forest management curriculum. You would graduate with a B.S. degree in forest management.

Other schools offer the forest management degree with various options plus separate degrees in some of the specializations. At these schools, it is possible to get a B.S. degree in forest engineering, in range management, or in many other areas.

Employment Prospects

Your job prospects will depend on a number of factors, including whether you major in forest management or in one of the specialized areas. The graduate with a forest management degree has the widest variety of employment opportunities, especially in the government sector. This is, however, an extremely popular field, and as such it remains very competitive. Succeeding in this specialty requires excellent grades, ambition, as much work experience as you can attain, and the ability to present yourself as intelligent and articulate to prospective employers. See Chapter 6 on employment opportunities for more information.

Forestry Education in the Twenty-First Century

The modern forester must adapt, integrate, and keep up with rapid and exciting advances in technology and scientific knowledge. The changing social and economic climate also calls for foresters to keep abreast of new demands and responsibilities. Naturally, the responsibilities of today's forester are quite different from the duties of a forester just ten years ago. And in another few years they will no doubt change again.

To meet these shifting challenges, forestry schools are continually updating their curricula. The tough question they must answer is not "What education does *today's* forester need?" but "What education will *tomorrow's* forester need?"

The skills a forester needs are increasing and changing faster than ever before. With the United States facing troubling economic times, foresters are being asked to pick up the slack and do more on smaller budgets. Foresters thus are engaging in a greater variety of tasks, from administrative duties to fieldwork. Nearly every forester mentions the need for thorough computer training for today's professionals. Modern foresters conduct surveys of forests using computer maps. Others analyze data from satellites. Still others may need to communicate using word processing or desktop publishing. In short, computer skills are indispensable for the modern forester, and most forestry schools make them an integral part of the curriculum.

A forester today also must be keenly aware of the growing environmental problems facing the world. Forestry scientists are on the leading edge of research into global warming, for example, studying how the "greenhouse effect" will change the shape of forests and ecosystems in the years to come. Increasing attention is being focused on making the harvesting of forests compatible with the needs of wildlife and other multiple uses. Consequently environmental issues are playing an ever-increasing role in the forestry curriculum.

Foresters employed by private industry are likely to specialize in traditional tasks such as timber inventory, regeneration, silviculture (development and care of forests), and logging. Public foresters, working for state or federal government, tend to specialize in planning and personnel management. Both public and private foresters enjoy a tremendous amount of diversity in their work.

Foresters agree that certain kinds of work are becoming more important to young, early-career foresters. They put silviculture, regeneration, program planning, computer analysis, interagency coordination, and public relations in this category. As you might expect, personnel management becomes a more important task as professionals gain experience and are promoted.

People often become foresters because they love the outdoors. However, once they attain some success at a professional level, continuing advancement often leads to more office work. So the "successful" foresters work themselves out of the woods into administrative jobs.

Whereas some foresters in the past may have worked in more remote areas with less public contact, today's foresters are dedicating more energy to people-oriented tasks. Planning, administration, public relations, and personnel management are as much a forester's duties as timber inventory. What is more, this trend will continue. As a result, some foresters believe that their education is lacking in these areas. They feel deficient in applied economics, mathematics, decision-making skills, and social sciences.

Many forestry schools have enhanced their curricula in these areas. As a forestry student, it will be worthwhile to make certain that your education includes lots of the above courses. Tomorrow's forester will have to be even more skilled at working with both people and trees.

Forest Engineering

Do you enjoy mathematics and solving mechanical problems? Then forest engineering might be the forestry specialization for you. Where other foresters manage the growing forest, the forest engineer is primarily concerned with harvesting the mature trees and getting them to market.

Job Duties

No two logging sites are the same. At each site, it is up to the forest engineer to determine the best way to get to the trees, cut them, and then load them onto the log trucks. The engineers design the systems of logging roads and oversee their construction. They also design the bridges, culverts, loading platforms, and other essential facilities.

All this must be done with an eye toward protecting the environment as well as the budget. The engineer's challenge is to devise an efficient and economical way to harvest trees while protecting the soil, water, wildlife, and other forest assets.

Education

If you are particularly interested in pursuing a career in forest engineering, be sure to select a forestry school with this degree option. Every graduate forester has a working knowledge of road design and logging site operation, and most forestry schools offer solid engineering courses, but there are certain forestry schools that provide a five-year program that will allow you to specialize in engineering. Upon completion of this demanding sequence, the graduate is awarded a bachelor's degree in forest engineering and a second degree in civil engineering. This means that the new forester can be registered as a licensed engineer as well as a forester.

You can also find plenty of programs that have an engineering emphasis within their regular four-year forest management programs. Many forestry school advisors suggest that incoming students sample a few engineering courses before throwing themselves into that option, which is good advice. There is plenty of time to specialize in the upper division and graduate years. But, if you think there is a forestry engineer hidden in you, make certain the school you choose has an engineering curriculum.

Employment Prospects

Forest engineering is a highly professional and specialized field. There will always be a demand for forest engineers, as timber companies and resource agencies cannot survive without their expertise. Although there will always be high-paying and responsible positions within this specialization, there will overall be a limited number of jobs in this area. Even the biggest timber company needs only a certain number of engineers, so the market can certainly become saturated with qualified applicants. Notwithstanding these practical considerations, a degree in engineering is generally considered a ticket to success. Most timber company executives expect the high demand for engineers to continue into the next decade or more.

Resource Recreation Management

Does the idea of majoring in "fun" sound intriguing to you? If so, you're in luck. One of the fastest-growing departments in many forestry schools is the resource recreation section because of the increasing demand for skilled recreation specialists. These foresters manage wooded lands to suit the needs of the hordes of outdoor enthusiasts.

Outdoor recreation has become an American obsession for several reasons. For one, Americans have more leisure time available than in the past. Improved gear has made outdoor sports accessible and attractive to nearly everyone. And the crowded conditions in our cities lead residents to flee at every opportunity to seek the peace and quiet of natural areas.

Job Duties

Forests are not simply places where trees grow. They are natural playgrounds for backpacking, fishing, hunting, camping, hiking, bird-watching, berry picking, and photography. The list of popular outdoor pursuits is growing.

On some of our public lands, parks for example, providing recreational opportunities is the primary mission. Here, the challenge is to create a sustainable balance between the needs of recreational enthusiasts and the natural resources themselves. With good management it is possible to provide for the optimal enjoyment of the forests by the greatest number of people. The challenge for some foresters is to manage the people and the property so that everyone can have fun without bumping into each other—and without degrading the natural value of the area.

On the other public lands, such as national forests, recreation is just one of many uses that must be accommodated. Here the manager must determine how to integrate all the leisure and sports activities with logging, mining, and wildlife management.

Timberlands owned by the wood industry are managed, of course, primarily to produce marketable logs. But even here recreation has become an important consideration. Many timber companies are allowing more and more recreation on their lands. Usually this is done as a public service, but it can also be a source of revenue.

Education

Outdoor recreation management is an ideal career for foresters who really enjoy working with people. The recreation manager must understand people's leisure-time needs. He or she must also have a good working knowledge of ecology, fiscal procedures, program design, government, and social trends.

There are a number of specialized options in recreation administration. Their names vary from school to school, but the areas of focus are generally these:

Outdoor Recreation

This specialty gives students a broad understanding of the relationships between recreational practices and natural resources. Planning skills are emphasized. Students learn how to develop a recreational site or facility and how to help people have a quality outdoor experience.

Graduates are employed by government agencies such as the Army Corps of Engineers, the Soil Conservation Service, the Bureau of Outdoor Recreation, and the Extension Service of the Department of Agriculture. State and county parks also need administrators. And some outdoor recreation experts are hired by private industry.

Park Administration

This program prepares individuals for employment with state, county, or federal park systems. Coursework concentrates on the planning skills necessary to design and operate public parks. The challenge is to provide maximum recreational opportunities with the minimum of environmental impact. Like other recreation managers, the park administrator must devise ways to reduce conflicts between users. Snowmobilers and cross-country skiers, for example, do not get along well together. Nor do bird-watchers and hunters. Swimmers have to be segregated from water-skiers. And the solitude seekers want to be away from everyone.

The modern park has modern problems. Vandalism, litter, pollution, and congestion—once identified as city problems—are now common in our parks. They all detract from the outdoor experience that draws people to the park in the first place. The park administrator must deal with these headaches while also protecting fragile natural values. Park users cannot be allowed to erode the soil, trample the vegetation, deface natural features, or force out the wildlife. As the challenges of resource management increase, more park administration specialists will have to be hired. It is no longer enough simply to set land aside for public enjoyment. Public places have to be managed by experts.

Outdoor Interpretation

This is the specialty to zero in on if you want to become a park ranger, camp counselor, or guide. Interpreters help and encourage the public to learn about the outdoors and to take pleasure in the experience. Through these educational efforts, they encourage an awareness of human impact on the environment. To do this, you will need not only a background in natural resources but good communication skills as well.

Forest recreation curricula are loaded with courses about people and about nature. Your classes as a recreation major might include psychology, education methods, natural history, museum methods, speech, economics, journalism, and environmental interpretation. These courses would be in addition to your basic science and forestry studies.

Graduates of this specialty may become park naturalists, directors of nature centers, or directors of interpretive programs for resource agencies. Administrators of botanical and zoological gardens, departments of archives and history, and historical sites also employ these specialists.

Some schools offer even narrower programs designed to train students for special interpretive jobs with social institutions such as prisons and orphanages, in commercial tourism and industry, and in inner-city parks.

Employment Prospects

Because of the increasing demand for quality outdoor recreation, there are lots of opportunities for professional resource recreation managers. But the curriculum has been very popular, and there often are more graduates than jobs available. In fact, many forestry school advisors have been discouraging students from enrolling in this specialty. They know that competition for jobs in recreation management is fierce. Still, an ambitious graduate with a glowing scholastic record should be able to find a satisfactory starting position. Summer work experience, as always, is quite valuable in helping you land that first full-time job.

Education and Communication

"Trees are easy to manage; people aren't." So goes an old saying among foresters. Forests and other ecosystems obey the age-old laws of nature. The effects of management techniques are statistically predictable. People, on the other hand, obey few natural laws and are seldom predictable. People present the forester with the most difficult challenges.

Foresters must make decisions that involve energy and water shortages, availability of outdoor recreation, jobs, local economics, and the quality of scenic vistas. These issues attract public attention. People today are involved and demanding. They have strong opinions and want to be heard. No forest management program is complete without the understanding and support of the local populace. That is why communication skills are so important to the forester.

The modern resource manager must be able to listen, to interpret, to inform, and to motivate. Many veteran foresters encourage today's forestry students to focus on acquiring strong communications skills and to get media training while in school. Without these assets, the task of getting the public's cooperation and support is that much more difficult.

Employment Prospects

Foresters with media expertise will find a growing marketplace for their skills. Such specialists are needed to build communication bridges between resource scientists and citizens.

Most forestry schools include communication courses in their general curricula. The student who specializes in this area, however, will stand out from the pack due to his or her in-depth training in journalism, broadcasting, media relations, education, speech, and public relations. Good communications skills will make a beginning forester more attractive to most employers. Many of the best positions for those who have specialized in communication and education will be found in government. Many publicly employed foresters, such as park rangers, find that once they begin working they become teachers first, foresters second. State and national forests, nature centers, and wildlife refuges need people who can communicate the conservation message to the public. Another source of employment for communication experts is found with some timber companies, especially those managing forests near urban centers. These media-wise foresters are expected to explain the industry's management policies to the public.

The expectation is that the need for public relations talent among foresters will remain strong. The employment outlook for these specialists should be good.

Range Management

Range is broadly defined as uncultivated land capable of supporting wild or domestic grazing animals. Using that sweeping definition, prairie, desert, tundra, meadowland, alpine mountain slopes, savanna, and wetlands are all range. In other words, range is natural grassland. It is distinguished from forest or cropland even though the three often run together in the same area.

Nearly half the landmass on earth can be classified as range. In the United States, there once were vast buffalo prairies that were lush stands of native grasses. Although most of what was prairie has been tamed and converted to other uses, there still are millions of acres of natural range in the western United States. Much of it is public land managed by federal agencies such as the Forest Service and the Bureau of Land Management.

Job Duties

Many forestry schools, especially in the West, offer majors in range management. Is a range manager a forester? Maybe not, but there are lots of similarities. Both work with large, wild ecosystems and employ the same ecological principles. Range management and forestry are both aimed at sustaining yield and allowing for a variety of uses. Both these professionals can be found working for government land management agencies or large land-holding companies. Similarly, both fields can lead to careers in education, research, or the private consulting industry.

Range is cow country. Much of the beef in the United States is grown on public range. Cattle and sheep ranchers apply to the local range manager for permits to let their stock graze on the range. The managers must ensure that overgrazing does not destroy the grassland. There are other considerations, too. Range is also home to mule deer, elk, antelope, bighorn sheep, upland game birds, and other wildlife. The manager must make sure that wild animals are not forced from their homes on the range by overgrazing livestock.

Range is also a valuable watershed. Water is a precious commodity in the West. Range managers are charged with ensuring that the water that runs off their grassland is of good quality. Range specialists also oversee the construction of water holes for livestock and wildlife. In many areas, they help design irrigation systems and monitor other water diversions.

Since most range is public land, it is heavily used by outdoor enthusiasts. The range manager, then, is responsible for a vast, open playground. Many people come just to enjoy the wide, open spaces, the natural beauty, and the clean air. Others visit the range to camp, hunt, hike, race off-road vehicles, fish, search for gemstones, or meditate. If the range is truly administered for multiple use, the manager will see that all these recreationists can enjoy their individual pastimes without diminishing the wildlife, livestock, or watershed values.

Range managers everywhere are now trying to deal with another visitor—the energy and mining companies. The western rangelands lie over huge deposits of oil, gas, and valuable minerals. The challenge is to extract those much-needed resources without permanently destroying the range for other uses. Range scientists are becoming increasingly active in the preparation of environmental impact statements and in energy exploration activities. Their special knowledge of grasses and shrubs gives them a leading role in devising plans to lessen the impact of mining activities. After the bulldozers leave, the range experts use their skills to help the area recover its former productivity.

The job description of the typical range manager has been broadened by modern circumstances. Whereas he or she was once occupied mostly with writing grazing permits, monitoring range conditions, and overseeing range improvements, the scope of work has expanded. The manager must now contend with a steady flow of questions and problems from a wide variety of sources. There are conflicts between range users to resolve and impacts to assess. In addition to the traditional grazing permit holders, there are motorcyclists, four-wheel-drive vehicle enthusiasts, hikers, horseback riders, energy and mineral prospectors, real estate dealers, and home developers who desire access to the range. The manager's challenge is to balance all these uses without allowing the combination to deplete the range.

Education

Students in range science must have a solid science background. Early courses in biology, soils, chemistry, math, computers, and physics will lay this groundwork. Then the student can delve into the more specialized courses. Upper-division courses can include the following:

Range watershed management
Range livestock production
Range economics
Range wildlife management
Grassland plants
Vegetation sampling methods
Forest and range relationships
Animal nutrition

Advanced courses in botany, zoology, soil science, forestry, and economics are also included.

Some schools allow students to narrow their majors in range management even further. Narrow specializations are offered in range economics, range watershed management, range livestock production, and forest-range management.

Employment Prospects

Americans are going to demand a lot out of their rangeland in the coming years: more beef, more wool, more minerals and energy, and more quality recreation opportunities. Meeting those demands will require the services of professional range scientists. The outlook for employment in this field is reasonably good. Although employment growth will not be as rapid as in some fields, retirements and the increased concern about the environment should create some openings.

Graduates from a range science program are qualified to apply for a variety of federal and state government positions. They can be forest rangers, soil conservationists, extension agents, range managers, or range conservationists. These are official titles used by such federal agencies as the Forest Service, Soil Conservation Service, Bureau of Indian Affairs, and Bureau of Land Management.

Range scientists also may seek employment in the private sector. Using their special knowledge of "applied range ecology," these graduates make good managers of large livestock ranches. They are also in demand as technical foremen for livestock companies, advisors to land management companies, and land appraisers.

Wildlife Management

Forests and wildlife just naturally go together. In the United States, the practices of forestry and wildlife management grew up together. The founder of the American science of wildlife management—Aldo Leopold—was a forester. In 1918 Leopold was supervisor of a national forest in northern New Mexico. He popularized the term *wilderness* and was one of the first to seek protection for it. He helped President Theodore Roosevelt introduce the concept of conservation to the public.

In 1933 Leopold published *Game Management,* a book that became the foundation of a new science. In this classic work, Leopold explained that wildlife is a crop, like corn. Nature produces a bumper crop of most game species every spring. Then nature, through environmental mortality factors such as predation and disease, takes away the surplus animals until the populations balance each other. We can have more or less of some wildlife species by manipulating environmental factors.

All wildlife needs food, water, shelter, and living space. An area that provides those essentials is called a *habitat.* The forest provides a perfect habitat for a wide variety of animals, so foresters frequently become involved in wildlife management.

Any given species requires a suitable habitat to survive and thrive. Our laws, good intentions, and environmental impact statements must be harnessed to provide protection for the proper habitat for each species. Part of the responsibilities of a forester charged with making multiple use of timberland is to understand the habitat needs of the resident wildlife and to use the existing laws to provide the necessary protection.

Trout and some other fish need clear, fast, cold water. If, during a logging operation, the forester fails to protect a trout stream from soil erosion and damming debris, the trout may not survive. Of course, there are lots of other reasons for guarding the quality of woodland streams, but if the fish die, that illustrates that the forester has not properly managed the logging operation.

Grizzly bears need lots of room and prefer wilderness. The wildlife specialist in grizzly country must decide which areas can be logged without disturbing the bear population. Spotted owls can live only in old-growth forests. But old-growth timber is very valuable. Foresters are often deeply involved in working to resolve exactly this sort of conflict, determining whether there are workable compromises that protect both the wildlife and the economic interests.

Ruffed grouse need fallen logs from which they can sit and call to their mates. The red-cockaded woodpecker, an endangered species, requires a certain size of pine tree for nesting. Kirtland's warblers, also endangered, can only reproduce in groves of young jack pines that have recently been burned over. The wildlife specialist should be able to identify these special needs and design a management plan to accommodate them.

Some wildlife species benefit from logging activities. Deer, for example, flourish in recently clear-cut areas. Harvesting the mature trees allows sunlight to reach the forest floor. Within a few months, grasses and annuals spring up and provide excellent deer food. After three or four years, the clear-cut is thick with young trees and shrubs, both food and cover for deer.

Job Duties

Managing the habitat resources for wildlife is frequently the responsibility of foresters employed by government agencies. Many, if not most, public forests are operated to maximize their wood, water, and wildlife values. Wildlife has tremendous recreational value and serves as an indicator of a forest's ecological health.

Foresters drawing their paychecks from industry are concerned primarily with wood production. Many commercial timber harvesters, however, are hiring foresters who are wildlife specialists. From the smallest private woodlot owner to the largest industrial giants, the trend is toward more concern for wildlife and other environmental factors. Law requires some of this concern, especially for endangered species. But many companies include wildlife in their management plans because it's good for their public image or simply because they actually are truly concerned about creating a sustainable ecosystem for the wildlife.

Even so, the professional forester soon realizes that, since dollars are important, wildlife must occasionally take a backseat to other forest assets. In harsh economic terms, the only income wildlife produces is through sale of access to the forest to sports enthusiasts. For this reason, wildlife is considered by many to be less important than timber production, agriculture, grazing, mining, water utilization, urban development, and other high-dollar uses of the forest.

Foresters must consider yet another aspect of wildlife: some wildlife threatens efficient wood production. From mice and porcupines to elk, many animals feed on young trees. Wildlife sometimes makes reforesting a logged area very difficult—and expensive. So the forester may become involved in protecting the trees from the animals.

The public is very interested in and concerned about wildlife issues. To manage wildlife effectively, the specialist in this field also must be able to manage people. The wildlife manager must be a skilled communicator, educator, politician, and community activist.

Every forester needs some knowledge of wildlife. It is an intrinsic part of the science. But the specialist must have years of special training. Wildlife management is a sophisticated science. The specialist counts animal populations, helps write and enforce game laws, studies food habits, controls animal damage, and protects endangered species. He or she works on habitat improvement programs, modifies logging plans, and, like most foresters, spends a great deal of time working with the public.

Education

Many forestry schools offer programs in wildlife science. This course of study is for the student who wants to be a forester with a wildlife specialty. Students who want to put the priority on wildlife should major in fisheries and wildlife science. Jobs for these majors include game biologists, refuge managers, administrators, teachers, and researchers.

The wildlife specialist must have an extensive background in science and ecology. Training in the social sciences, communications, economics, and government is also essential. The forester emphasizing wildlife management will take all the basic forestry courses. Upper-division wildlife course work will include the following:

Principles of wildlife management
Population ecology
Ichthyology (study of fish)
Ornithology (study of birds)
Limnology (study of lakes and streams)
Mammalogy (study of mammals)
Wildlife law
Wildlife census techniques
Animal nutrition
Anatomy and physiology

The well-rounded curriculum will include intensive studies in government, media use, statistics, and social sciences.

Some schools offer two-year programs that lead to wildlife technician degrees. These programs give the technician the skills necessary to work in biological or agricultural laboratories. Technicians also learn to collect biological data and to perform technical tasks around a wildlife refuge or hatchery.

Employment Prospects

Because this is a particularly appealing specialty to many forestry students, the competition for positions in wildlife management is quite fierce. Graduates with double specialties have an edge, as do those with advanced degrees. As usual, good grades, ambition, summer work experience, and enthusiasm are assets that will help score job offers.

Urban Forestry

Three out of every four Americans live in a city. Trees help make cities pleasant places to live. They provide shade, greenery, flowers, oxygen, and a refreshing touch of nature. In backyards they serve as a stopover for wildlife and a place to hang a swing or a hammock.

For both practical and aesthetic reasons, trees are important to those of us who live in increasingly crowded cities. The management of trees in the "asphalt jungle" presents special problems. The trees have to tolerate such hazards as traffic, polluted air, children imitating Tarzan, and sidewalk-covered roots. Likewise, people want "civilized" trees that do not shed too many leaves, drop branches on unsuspecting heads, topple in slight windstorms, or buckle sidewalks with unruly roots.

Job Duties

Urban foresters help plan parks, greenbelts, and highway landscapes. Special problems may be as small as deciding which trees must be cut for a power line, or as important as combating dire threats to urban trees such as Dutch elm disease or the more recent menace of the Asian Longhorn Beetle. Trees may rank among a city's prize possessions. It is the urban forester's job to preserve these bits of living heritage.

Education

A course of study for urban forestry would emphasize the physical, social, and political systems of cities. Classes in planning and communication would be important additions to the broad forest science background needed to meet the challenges of managing inner-city forests.

Students interested in pursuing urban forestry should contact the National Urban and Community Forestry Leaders Council. This group can be reached through either the Forest Service or the American Forestry Association. There are members of the council in most major U.S. cities, so it would be possible for a student to visit an urban forester and see the work firsthand.

Employment Prospects

Municipalities of all sizes hire urban foresters. They may be part of the parks board, attached to the planning department, or integrated elsewhere into a city government. Some urban foresters work for universities or for state extension services. City forestry is an excellent specialty for the forester who wants to begin a private consulting business. (See Chapter 6 for more information on urban forestry.)

Recent decades have seen an even greater shift of population to cities. With that trend came the realization that a healthy urban environment is critical. Therefore, jobs in urban forestry are flourishing—although they are subject to the difficulties many cities are having raising money. Overall, however, this is an excellent opportunity for current forestry graduates.

International Forestry

More than a third of the world's land is either covered by forest or is capable of supporting it. As the world's population experiences explosive growth—exceeding six billion people at the start of the twenty-first century—the demand for wood intensifies. Yet forestland is one of the basic ecosystems that ecologists consider threatened by human exploitation and waste. There is a critical need for trained forest managers in countries around the world.

In parts of India and Africa, the energy crisis is a wood crisis. Firewood, the principal fuel, has become dangerously scarce. In South America, the lush jungle is being cleared at an alarming rate to make room for cattle and crops. In situations like these, the erosion and depletion of the natural resources creates even greater environmental pressures and problems. In many other parts of the world, demand for forest products is outpacing yield.

Education

An international forestry curriculum will prepare you for a job abroad. Your classes will concentrate on the physical similarities and differences among the forests of the world. There will be emphasis on the cultural and social factors that affect the way forests are managed in different countries. You will also study international trade mechanisms to get an understanding of the world market.

Work in the international field requires special tools and attitudes. Foreign language skills are vital. A forester expecting to accomplish anything in a foreign country must be adaptable and sensitive to cultural differences. And here again, solid managerial abilities are extremely useful, and they will be put to the test in possibly very unique circumstances.

Students who elect to study international forestry typically seek advanced education in some special managerial field. This combination makes them attractive to employers both at home and abroad.

Employment Prospects

Many of the largest timber companies operate in foreign countries and maintain branch offices overseas. Foresters with international qualifications are also needed by foreign aid agencies such as the World Bank and the United Nations Food and Agriculture Organization. For the young graduate, the Peace Corps probably offers the most immediate opportunity. There are also exciting possibilities with nongovernmental organizations as well as working with research and university scientists.

Forest Products Industry

One of the most exciting and rewarding fields of forestry work may keep the forester out of the woods altogether. Instead, the forest products specialist may work in a brightly lit laboratory or a modern paper mill. These science-minded businesspeople also find positions in sales, personnel management, administration, teaching, and financial control.

The outlook for wood and fiber science specialists is bright. In fact, foresters with this background will likely find jobs more readily than resource managers.

Most forestry schools offer a forest products program. Options for concentrated study include wood products, which emphasizes production, sales, and technical services. The wood science option stresses science and technology in wood and bark utilization. This course provides a good base for advanced degree work for students interested in research, product development, and teaching. A third course concentration, pulp and paper technology, prepares students for employment in the progressive paper products industry. The terminology varies from school to school, but these generally are the three specialties within a forest products curriculum.

Wood Products

Wood is literally a building block of civilization. It is not only the most diverse building material; it is the only material that is renewable. Other metal construction materials are exhaustible and increasingly expensive. So wood, the most ancient construction material, is also the material of the future.

Tree parts are used for much more than building materials. More than five thousand different products have some tree in them. Wood, or its by-products, can be found in everything from plastics to explosives, from bags to bubblegum, and from the cheapest cattle feed to priceless violins. Forest products specialists are finding new ways to use wood every day. They are also devising methods to improve products and to use every tree more efficiently.

After a tree is harvested and hauled to the mill, its versatile fibers may be processed in many different ways. The wood technologist is familiar with the unique properties of the different kinds of wood. Oak, for example, is dense and strong, useful where support is needed. Although pine is not as strong, its long fibers are perfect for paper production. Black walnut does not make very good paper, but its strength, rich color, and beautiful grain patterns make exquisite furniture. The challenge is to make the best use of the special properties of each variety of trees.

Opportunities in manufacturing will test a wood technologist's inventiveness and his or her knowledge of engineering, physics, chemistry, and mathematics. Processes are devised to get the full potential out of each tree. Manufacturing specialists must know how a tree is built so they can take it apart, fiber by fiber. Then they draw on their training and imagination to recombine the fibers into a new, more useful product.

In other cases, wood is not dissolved by chemicals, shredded, or otherwise taken apart. It can be sawed, shaped, heated, bent, pressed, glued, treated, and painted. Eventually the wood is made into products as diverse as fine furniture, toys, and spaceship parts.

In every process the goal is to take advantage of the wood's desirable qualities. Different woods offer the product designer varying degrees of durability, flexibility, strength, beauty, and so on.

Modern manufacturing plants need a cadre of workers with a variety of talents. Engineers are needed, of course, as are chemists, electricians, designers, drafters, quality control agents, efficiency experts, equipment operators, mechanics, and an army of skilled laborers. In a forest products plant, all these people have something in common—they all have knowledge of wood. Even those not involved in the hands-on manufacturing process—the purchasing agents, personnel managers, accountants, and statisticians—must understand the forest products industry. With a college degree in forest products science and the appropriate specialization, you could fit into any of these slots.

Marketing and Distributing Wood Products

One can imagine meeting a forester in a remote national forest, but you might also encounter one on Madison Avenue. Foresters are no strangers to the marketing centers of the world. Many of them have specialized education in the promotion and economics of forest products. Marketing is another bright facet of the forest products industry.

To sell a product as highly technical as a wood derivative, you have to know all about it. The successful salesperson knows how the product is made, its special qualities, and its advantages over the competition. For the forest products marketer, this requires a technical background, usually including a knowledge of forest management.

Students interested in forest products and business can find employment as salespeople, economists, advertising specialists, publicists, designers, artists, and marketing managers. Worldwide expansion of the forest products industry has opened up opportunities in foreign trade, overseas marketing, transportation, and foreign business planning.

If you are comfortable working with people, economics, and trees, a focus on forest products marketing might be for you.

Education

To prepare for a career in the wood products industry, you will need a solid background in the physical and biological sciences as well as specialized courses in wood science. The courses in your curriculum will depend on your major and which school you attend. Some programs are mostly business and management courses. Others are loaded with organic chemistry, computer science, and math courses.

Some specialized wood science courses include:

- Wood frame construction materials
- Commercial trees of North America
- Wood structure
- Wood mechanics
- Chemistry of cellulose and wood
- Wood seasoning and preservation
- Timber harvesting
- Mill operation and inspection
- Composite materials
- Evaluation of production systems

Most wood science or wood technology programs also include studies in basic forest management and a generous number of liberal arts courses.

Employment Prospects

The outlook for wood and fiber science specialists is promising. In fact, foresters with this background will likely find jobs much more available than will resource managers. Resource management jobs are usually limited in number because there are only so many forests to manage, and competition for these coveted positions is often intense. But jobs developing and marketing new wood products are limited only by the ingenuity of those in the field. Fortunately, inventiveness has never been in short supply among wood technologists.

Wood Science

Humankind has used trees since the very beginning, yet we haven't stopped developing innovative ways to use wood. Researchers working closely with forest products manufacturers are constantly devising new and improved methods and products. These specialists are usually graduates of a forest products science school. Often they have additional education in chemistry or physics.

Why get involved in wood research? It seems a reasonable question. Since we have been studying wood for centuries, isn't the field pretty well exhausted? It is not. In fact, forest products researchers are opening exciting new doors every day. The potential of wood has never been greater, and the research opportunities have never been more intriguing.

The cost of energy will continue to rise. Since processing wood requires a lot of energy, researchers are looking for ways to make operations more energy efficient. Some processing plants are already using the sun to cure newly cut hardwoods in ultramodern solar dryers. Researchers are exploring solar and other alternative energy sources.

The imaginative use of space-age adhesives has made dozens of new products possible. Wood chips and scraps are glued together to make versatile construction panels. Wood is being bonded to dozens of other materials to create all sorts of useful combinations. Researchers are looking for new adhesives that do not require expensive petroleum.

A whole catalog of chemicals can be extracted from wood. Turpentine, alcohol, pine oil, flavors, fragrances, and industrial chemicals are just some of wood's by-products. Two challenges facing researchers are how to reduce the cost of extraction and how to get even more useful products from materials now wasted. At the research lab, we can watch scientists use fungi and microorganisms to convert wood to chemicals.

Another exciting area of research lies in the development of affordable housing. Population pressures have resulted in increasing new home construction. Forest products researchers are helping to meet this need by developing structures that are stronger and yet use less wood. They also are looking for ways to make homes more energy-efficient and fireproof.

Scientists are also looking for better ways to protect wood from its old enemies: fire, insects, weather, and decay. Other foresters are studying foreign woods from all over the world. Some of these exotic trees may prove valuable to our domestic wood needs. These are some of the research projects underway at just one laboratory. As you can see, there are lots of fascinating problems in the wood products area just waiting for bright young scientists.

Another forest products company learned to compress wood scraps into pellets that can be burned like coal. Wood can be converted into both natural gas and that international energy staple, oil. And don't forget firewood! As energy prices continue to rise, more and more home owners are turning to this old standby. Researchers have developed stoves that produce more heat with less wood. In fact, wood, the oldest fuel, now heats more homes in America than one of our newer inventions, nuclear power.

Education

Researchers in wood technology usually have advanced degrees. Frequently, their B.S. will be in a physical science such as chemistry or physics. Their graduate degrees will be in wood science.

Employment opportunities are so diverse that a variety of degree combinations may lead to a research job. For example, the dean of a leading forestry school reports that one of his recent graduates with a degree in forest products and a second degree in chemistry is now researching wood preservation methods for a large construction firm. Another graduate with dual degrees in wood science and music is studying the tonal qualities of wood for a company that manufactures musical instruments.

Employment Prospects

Wood scientists can work in any of hundreds of different industries, government agencies, schools, research groups, consulting firms, and trade or professional organizations. The talented investigator with an exemplary scholastic record should have no trouble finding a job.

Forest products will play a leading role in our future. As a researcher in the field, you could enjoy the challenge and the rewards of matching the needs of society with the amazing variety of qualities offered by wood.

Pulp and Paper

Fully half of the trees harvested in the United States are chipped, shredded, cooked, and made into paper. Regardless of which forestry specialty you pursue, your career will probably be affected by the paper industry. And the paper industry as a whole is thriving. An ancient art, papermaking is also a developing, dynamic science. From forest to fiber to finished paper product, it is a field full of challenge and opportunity.

Paper was a catalyst for civilization. It allowed ideas to be written down, preserved, and transferred. Our schools, businesses, and bureaucracies run on paper. Paper wraps and bags everything; serves as money; brings our bills, letters, and news; and stifles a sneeze. There are more than one hundred thousand paper products. Paper is so much a part of our lives that we tend to forget about it.

Thanks to the work of foresters and paper technologists, North Americans have plenty of paper. Americans each use over six hundred pounds of paper every year, compared to about three hundred pounds used by each Canadian. Each person in China uses less than ten pounds.

Why, you might ask, are trees used to make paper? Tear any piece of paper, and hold it up to the light. See those hairlike fibers? Those fibers are the raw material of all paper. Papyrus, the first paper, was made by pressing stems of aquatic plants into sheets. Later, paper was made from the fibers of cloth. Fine "rag" papers are still made today, but they are very expensive.

The best fiber that can be obtained in abundance comes from the cellulose of trees. Any piece of wood can be used. Loggers send the best logs to sawmills to be cut into lumber. But the paper mill can use all the scraps, limbs, small logs, and even the sawdust that is left over.

The papermaking process usually begins with wood chips. Different kinds of wood chips are mixed in huge vats according to a recipe. There they are treated with chemicals and cooked until they break down into a slurry called "pulp." This mush is processed, pressed, and dried. The result is paper—newsprint, fine writing paper, containerboard, or specialty paper, depending on which recipe is used.

Graduates of a wood and fiber science school are active in this process from start to finish. They grade the wood chips coming into the mill, determine the mixture of heat and chemicals applied, and oversee the production. Other paper specialists are involved in quality control, management, and customer service.

The paper industry's research departments are very active. Scientists are searching for ways to cut energy consumption in the paper mill. Others are adapting the latest in computer controls to the production process. Imaginative specialists are inventing new papers that are lighter, stronger, and more resistant to age. Environmental engineers are hard at work reducing the mill's pollution output.

Education

In addition to basic forest management courses, the paper specialist takes courses such as the following:

Introduction to papermaking
Pulp and paper processes
Chemical engineering
Physical chemistry
Wood and fiber characteristics
Wood chemistry
Chemical and physical properties of paper

Paper coating
Paper machine operation
Principles of management
Air and water pollution engineering

Employment Prospects

In the past fifty years, the per capita use of paper in the United States has increased greatly. The explosion of computer technology has not diminished this need. With computer printers, fax machines, and advanced copiers spewing forth more paper than ever, predictions of a "paperless" society have not come true. Production of paper products is expected to go up in coming decades. Those expectations assure newly graduated pulp and paper specialists that they are entering a dynamic industry with a promising future. Their challenge will be to use energy and wood resources more efficiently and to continually provide new and better paper products. This challenge is compounded by the need to meet these goals in an environmentally sound manner. To the ambitious young forest products graduate, this challenge adds up to positive career opportunities.

Forestry Technician

Attending a four- or six-year program in forestry is not for everyone. Due to time or financial constraints, many people choose not to pursue a regular university degree. Instead they choose to enter a program that will give them the training and qualifications to become a forestry technician. With their certificate or associate degree in hand, they are able to continue as technical workers in the field of their choice.

Dozens of "ranger schools" around the country offer one-, two-, and three-year programs that enable the graduate to enter the forestry field as a technical worker. Perhaps the most common is the two-year program leading to an associate degree. Some of these programs can be taken at night in conjunction with a full-time job.

In 1912 the State University of New York College founded the first ranger school. The campus is in Wanakena in the western Adirondack Mountains. This school, and most of the others like it, has commercial forests nearby that serve as outdoor classrooms. There are about sixty-five schools with forestry technician programs.

Entrance requirements to ranger schools vary. Most demand a high school diploma. Passing the high school graduation equivalency examination will usually serve in lieu of a diploma.

Job Duties

A technician's job description will vary depending on the employer, the part of the country, the season, and the abilities of the individual. Technicians are active members of the management team. They do much of the technical hands-on work, leaving the professional foresters more time to plan and administer. Most of the work is outdoors, on location, regardless of the season.

Technicians handle many of the important inventory duties. They mark trees for cutting, measure and calculate timber volumes, and record the data from timber sales. Technicians survey and mark the boundaries of areas to be logged or improved. Depending on their level of skill and experience, technicians can design and lay out logging roads or fire trails.

There are land leases to inspect and right-of-way agreements to record. These are a trained technician's duties. Those hours in the classroom come in handy when it is time to interpret aerial photos or map areas to be harvested or planted. A forestry technician usually heads tree-replanting crews. He or she will take the soil samples necessary to determine whether an area needs additional fertilizer.

Forestry technicians guard the timberland from insects, disease, and fire. They survey the area frequently, always alert for an infestation of pests or a fire hazard. If spraying is necessary, a technician may direct the operation. During the critical fire season, technicians can expect some long hours of lookout duty. If fire does break out, technicians will be on the front line. Their knowledge of fire suppression methods makes them valuable in the battle to save the forest. After the fire, a technician may be assigned to map the damaged area and estimate the damage.

Technicians often are responsible for managing their employer's buildings, facilities, and equipment. They make repairs, manage stocks, and see that work crews are properly equipped. Some agencies and many companies have nurseries and greenhouses. Technicians often staff these facilities. Other technicians are in information positions, explaining company or agency policy to the public.

Education

The typical two-year ranger school is an intensive crash course in forest management methodology. You will learn many of the skills taught in the longer forestry school curricula, but generally with less emphasis on theory and planning. As a student technician, you will get a sweeping overview of basic sciences and ecology.

Most of the class work is hands-on. Much of the schooling is done outdoors in real forest management situations. This practical training is good since employers want to hire technicians they can put right to work. Ranger schools aim to train technicians so that they will become quickly effective on the job.

Some of the specialized courses taught at forestry technician schools include:

Forest surveying and mapping
Road design and layout
Tree identification
Silviculture (development and care of forests)
Timber cruising (forest inventory)
Ecology
Water resources management
Forest protection from fire, disease, and insects

Employment Prospects

Forestry technicians are found working within every major government agency and private company that manages timberland. As increased demand for wood products forces landowners to intensify management practices, there will be a corresponding need for people with those special skills. Competition for jobs will be tough, however, with more applicants than positions available.

As a paraprofessional, the technician is an employer's compromise between a laborer and a certified forester. The technician earns less money than the professional and has fewer opportunities to climb the corporate ladder. Technicians can, however, win promotions to supervisory positions. They are paid reasonably well for their knowledge and experience.

After a few years of on-the-job training, the technician may be able to move up into a professional forester position, but this usually requires returning to college for additional training or to satisfy employers' requirements for a bachelor's or graduate degree.

6

EMPLOYERS OF FORESTERS

ARE YOU DRAWN to a career in forestry because you've met or worked with a forester from the U.S. Forest Service? Up until the 1950s, the majority of foresters worked for the Forest Service, and even today more foresters work for the Forest Service than for any other agency or company, and their work has a tremendous influence on those entering this profession. Their combined efforts put the Forest Service at the vanguard of international forestry. The Forest Service focuses on a combination of conservation, smart utilization of resources, and the management of wildlands.

The U.S. Forest Service

The Forest Service, which was founded in 1905, is part of the U.S. Department of Agriculture. It is responsible for managing more than fifty-one thousand square miles of the nation's 156 national forests, 19 national grasslands, and 17 land utilization projects. Altogether these lands encompass almost 191 million acres in forty-four states, Puerto Rico, and the Virgin Islands. This far-flung public domain contains an incredible wealth of natural resources. Precious water, wildlife habitat, forage, wood, recreational opportunities, and minerals are nature's gifts that the Forest Service manages. The service administers these resources under the concept of multiple use, striving for the greatest benefit to the greatest number of people over the long run.

The Forest Service has more than thirty thousand employees. Professional foresters make up more than five thousand of these employees. About three-quarters of the foresters work in national forests. The other foresters are engaged in administrative duties, education, or research at regional offices, experiment stations, laboratories, and other facilities. There are more than one hundred such installations across the country.

Forest Service activities are classified as national forest system management, cooperation, or research. Management means protecting and administering the 191-million-acre system under the guidelines of multiple use and in a sustainable and environmentally sound manner. Cooperation refers to the assistance provided to state and private foresters. The service not only manages its own lands with the most modern techniques; it helps other timberland owners do the same. Experienced foresters from the service share their expertise with state foresters, private woodlot owners, wood processors, private agencies, and other government institutions. The service operates nine major forest research facilities. In addition, it supports individual research projects from coast to coast.

Foresters

Foresters working for the Forest Service will be found engaged in the widest possible range of disciplines and activities. As a whole, most of the foresters are graduates of four-year forestry schools, having earned at least a bachelor's degree. The majority work as uniformed foresters in charge of woodlands. Here are some examples of the wide variety of projects that our nation's foresters are engaged in at any given time:

- Gathering the necessary data to prepare a resource management plan
- Planning and supervising the installation of flood control, soil conservation, and watershed improvements
- Managing wildlife habitat for big and small game as well as prime fishing streams
- Speaking with inner-city schoolchildren about conservation
- Maintaining recreational facilities such as campgrounds, trails, and swimming areas
- Measuring density of flammable materials in a forest to monitor fire danger
- Meeting with community economic development leaders

Your precise responsibilities as a forester would vary from site to site, but most positions call for a challenging combination of civic and scientific responsibility. As an up-and-coming forester, you often will have the opportunity to work in different parts of the country. An experienced forester with a distinguished record is a candidate to take over a ranger district. The next step might be to administer an entire national forest. Because the Forest Service is such a large, decentralized agency, the possibilities for advancement are good.

Other Professional Opportunities

With a forestry school degree that is slanted toward one of the other disciplines, you could put your training to work in the Forest Service. Or, if you have a related natural resource management degree, the service may have an opportunity that fits. There are many professional forestry jobs in the service besides forest management. Here is a sample:

Engineering

The Forest Service employs a variety of engineers. Most of them are civil engineers, of which there are approximately eleven hundred. Many of these have graduated from a forestry school's engineering program. The others are agricultural, electrical, industrial, and mechanical engineers. A degree from a four-year professional engineering program is generally required for an entry-level position. Advanced degrees will qualify candidates for employment at higher levels.

Geology

To scientifically manage natural resources, it is important to have a knowledge of the local geology. For this reason, the Forest Service has always employed geologists to work with land management projects. These days there is even more energy and mineral exploration of public lands, so geologists are in even greater demand. Geologists help design watershed improvements, bridges, dams, and roads. A geology degree is required for such work.

Hydrology

Public forest and rangelands occupy some of the nation's most important water-producing areas. The hydrologist is an important member of the management team on these lands. This water expert determines the influence of grazing, timber harvest, mining, road construction, and other activities on the area's ability to produce unpolluted water. Any one of a variety of science degrees will qualify you for this position. Suggested curricula are watershed management, hydrology, or aquatic biology.

Landscape Architecture

The national forests are America's backyard. Forest Service landscape architects work to make our land more usable and enjoyable. They assist in planning, determining location, and designing of facilities. Fire control stations, recreation facilities, ranger stations, and roads are all influenced during construction by the artful hands of landscape architects. A landscape architecture degree is required for such work.

Range Conservation

National forests and grasslands include more than one hundred million acres of range. These grazing lands provide forage for big-game animals and livestock. The range conservationist's challenge is to maximize production while protecting the range from overuse. Most four-year forestry degrees will qualify you for work on a range or grassland.

Research

Until recently most Forest Service researchers were graduates of forestry schools. Today most new research employees have received specialized graduate training in either a basic science or engineering instead of in forestry. Dozens of different scientific specialties are represented among Forest Service researchers. Except for the Forest Service research centers, most of the service's research facilities are located on college campuses. At least a bachelor's degree is required to get in the door.

Soil Science

Resource management programs, like trees, are rooted in the soil. An understanding of the soil is a prerequisite to developing management plans. Although trees can grow back, soil cannot. Soil scientists are needed by the Forest Service to oversee the protection of the soil. A four-year forestry degree with a soil science emphasis is an excellent combination for this option.

Wildlife Management

The Forest Service manages lands that are public property, seeing that the uses fit together into a comprehensive plan. One important element of this is protecting the natural habitats of wildlife. The national forests are home to about forty species of rare or endangered wildlife. They also harbor game populations and provide opportunities for some of the best hunting and fishing available in the country. This wildlife can be maintained by caring for the habitat. Biologists contribute to management plans and monitor wildlife populations. A forestry degree with a wildlife major can qualify you for this job, but those with a straight wildlife science degree may have the edge.

Technical and Support Positions

The professional forester may have the name recognition and some of the glamour, but behind the scenes are technical and support personnel who work in just as varied and exciting an environment. In fact, for every forester in the Forest Service, there are two or three workers in supporting roles such as technicians, aides, skilled workers, clerical personnel, and laborers. They are active and important members of the forest management team. For the most part, they do not have university degrees in forestry. This variety of support positions offers the high school graduate, or any ambitious person, an opportunity to participate in forestry without having to go to college.

The Forest Service classifies these support positions as follows:

- **Forestry aides and technicians.** Aides and "techs" perform such tasks as measuring stands of trees, recording data from weather stations, selectively cutting trees, maintaining campgrounds, and fighting fires.

- **Hydrologic aides and technicians.** These workers help the hydrologists in the various duties necessary to properly manage the watershed.

- **Physical science aides and technicians.** Operating instruments, mixing solutions, performing chemical analyses, and processing data are some of the jobs in this classification.

- **Biological aides and technicians.** These assistants work in biological and agricultural laboratories such as greenhouses, hatcheries, and wildlife refuges.

- **Engineering aides and technicians.** A basic knowledge of surveying tools and compass techniques is required to win this job. These workers help survey and design the many roads, trails, and facilities needed.

- **Surveying aides and technicians.** As part of a survey crew, these aides do the construction staking, chaining, clearing, and data keeping necessary to map an area.

- **Range aides and technicians.** This work is varied, but the assignment is to assist the range scientist in managing a grazing area.

A range aide might run a survey to measure the impact of grazing, fix fences, or construct a cattle-watering facility.

- **Trade and craft workers**. Examples in this category include carpenters, parachute packers, welders, cooks, bulldozer operators, and other skilled workers.

- **Forest workers/laborers**. Some of the duties of these laborers are pruning or culling trees, building fire lines, planting seedlings, and moving equipment.

- **Clerical personnel**. The Forest Service has paperwork, lots of it, in fact. The clerical worker's domain is the office. Receptionists and telephone operators have important jobs because they represent the Forest Service to the public.

- **Computer aides and technicians**. Computers are an indispensable tool in modern forestry. Skilled computer technicians are greatly sought after.

Qualifications for Support Positions

Applicants must be at least eighteen years old. When hiring, the U.S. federal government gives strong priority to U.S. citizens, and all persons must be considered without regard to race, sex, creed, age, color, national origin, religion, or disability.

Technicians are a grade or two higher than aides in pay and responsibility. A minimum of two years of experience is required for a tech position. High school or other education can be substituted for part of the experience requirement.

Depending on the level of the job, clerks need at least a high school diploma, and they often also need to have one or two years of work or post–high school education, such as at a technical or clerical college. Trade and craft workers need six months of experience or six months of trade school. Laborers must be in good physical condition. There are no education or experience requirements for laborers.

These are the minimum requirements. Any education or experience a candidate has beyond these minimums will give that applicant an advantage in the hiring process. Education or experience also may qualify applicants to enter at higher levels. A list of forestry technician schools is included in Appendix C.

Natural Resources Conservation Service

Several other agencies in the USDA besides the Forest Service hire foresters. The best known of these is the Natural Resources Conservation Service (NRCS), formerly known as the Soil Conservation Service. The agency was a product of the Dust Bowl of the 1930s. During those Depression years, poor farming practices and drought brought disastrous results. Precious topsoil from millions of acres of farmland was lost to wind and water erosion. The Soil Conservation Service, as it was called until 1994, was part of the federal effort to stop this frightening waste.

The NRCS is administered through local conservation districts. These rural offices provide technical assistance to landowners interested in protecting their soil. Depending on the needs of the region, a conservation district may hire soil scientists, range conservationists, hydrologists, or foresters.

These scientists advise landowners on watershed projects, windbreaks, flood protection, reservoir construction, and wildlife habitat. By surveying the farmer's soil, the NRCS technicians determine what crops would best grow there. Using this basic information, the scientist and the farmer draw up a land conservation plan for the farm. The object of this plan, of course, is to get the most productivity out of every acre while protecting the long-range value of the farm or ranch. While the landowner is putting this plan into practice, the NRCS advisors continue to provide the latest resource management information as discovered by research or experience in other areas.

A job with the NRCS is for those foresters and other resource specialists who really enjoy working with people at the grassroots. The local NRCS specialist is a teacher, a rural diplomat, and a scientist who stays abreast of modern management methods.

Cooperative State Research, Education, and Extension Service

The Cooperative State Research, Education, and Extension Service (CSREES) is the educational arm of the USDA. There is an extension office at each state land-grant university, and more than three thousand local offices. County extension agents are located in nearly every county in the country. These teachers work with their resident communities, providing technical information and support on everything from making honey to storing hay. Their students include farmers, gardeners, livestock owners, 4-H club members, and anyone else in need of such information.

Every extension agent is an emissary of the USDA's many research programs—an ambassador of information. Forestry management techniques are an important part of the intelligence distributed by extension agents. A forestry specialist with the county extension office may write a brochure on tree pruning in the morning, advise a woodlot owner on production in the afternoon, and, after dinner, speak to a 4-H club on Christmas-tree farming.

Department of the Interior

The Department of the Interior is the nation's principal conservation agency. Its mission is to protect America's natural treasures, while supervising the responsible and wise use of energy and mineral resources. The Department of the Interior (known as DOI) conserves and protects our water, fish, wildlife, mineral, land, and recreational resources. It also is the government agency charged with honoring our nation's commitments to Native Americans and Alaskan Natives.

The Department of the Interior is made up of many smaller divisions. Those of particular interest to foresters are the Bureau of Land Management (BLM), the National Park Service (NPS), United States Fish and Wildlife Service (FWS), and the Bureau of Indian Affairs (BIA). These agencies are major employers of foresters and forestry technicians.

Bureau of Land Management

The BLM has management responsibility for 262 million acres, more property than any other federal agency. In fact, the land managed by the BLM encompasses about one-eighth of the land in the United States! Most of this land is in the western part of the nation and in Alaska. These vast expanses include grasslands, mountains, arctic tundra, and deserts. In addition, the BLM also manages about 700 million acres of subsurface minerals.

Like the other agencies in the Department of the Interior, the BLM manages its public land under the multiple-use guidelines. This means it tries to maximize all the potential values of the land. These values include forests for wood; range for grazing; minerals such as coal, oil, and gas for energy; recreation; wildlife; and soil and water. The BLM also oversees wild horse and burro populations, wildlife habitats, and archaeological, paleontological, and historical sites. Balancing all these assets and uses demands the skills of professionals.

The BLM hires more professional foresters than any other branch of the DOI. The duties of BLM foresters vary according to location. Generally, though, their day-to day activities are similar to forest managers for industry or the Forest Service. The BLM sells timber from public land, so the BLM forester designs roads, conducts surveys, cruises timber, appraises blocks of timber, and administers sales. He or she may also work up plans for logging, slash (timber harvest residue) disposal, reforestation, and other silvicultural practices. Other BLM foresters may spend all of their time in forest protection or in administration.

Due to budget cuts, fewer recent college graduates are hired each year to join the approximately nine thousand current BLM employees. Those that come on board are usually employed at the midlevel federal grades of GS-5 through GS-7. The senior-grade jobs ranked GS-9 and above usually are filled by promoting professionals from within the bureau. Check with your nearest Federal Job Information Center to find out the current pay schedule for these grades. (See also the section on applying for a job with the federal government in Chapter 9.)

Most of the new professionals are hired to practice forestry, wildlife management, range management, soils science, minerals engineering, and realty classification. Support positions with the BLM include technicians and aides in all of the above categories. Firefighters, watershed workers, surveyor's aides, and clerical workers all find jobs in the BLM.

National Park Service

Congress created the National Park Service (NPS) in 1916 to protect, preserve, manage, and enhance our national parks. The parks include important natural, cultural, historical, and recreational areas. The national park system is made up of 385 park areas covering more than eighty-four million acres in forty-nine states, the District of Columbia, Guam, American Samoa, Puerto Rico, Saipan, and the Virgin Islands. Since the parks are so diverse—from beaches to battlefields and monuments to mountains—the NPS employs a full range of professionals and technicians to manage and administer them.

Forestry students often contact the NPS asking about job possibilities. The Park Service does not employ many professional foresters. There are, however, some jobs in the NPS that may interest the forestry student. Rangers, natural resource managers, engineers, and planners are among the possibilities for those interested in working for the NPS.

The Park Ranger

The park ranger, wearing a government green uniform and flat-brimmed hat, is often mistakenly called a forest ranger. Forest rangers work for the Forest Service managing national forests. Park rangers work for the NPS managing national parks. A forestry degree or forest management experience is required to get a job as a forest ranger. A forestry degree or experience is not required for employment as a park ranger.

Park rangers plan and carry out conservation efforts. They protect the park's plant and animal life from fire, disease, and overuse by visitors. They plan and conduct programs of public safety, including law enforcement and rescue work.

Most park rangers are involved in interpretive activities. They help the park visitors enjoy and learn from the park by conducting guided tours, presenting slide shows, or even performing dramatic reenactments of historical events. Rangers produce environmental education programs to teach park-goers, especially schoolchildren, about the relationships between people and nature. Park rangers should enjoy public speaking because they must often give talks to groups both in the park and in the local community.

Much of the park ranger's work is outdoors, but there are plenty of office chores, too. Rangers plan recreation activities, conservation programs, and park organization. They also do some budgeting and accounting. As rangers get more experience, they should be promoted into supervisory and administrative positions. That means more time in the office and less time outdoors. Many of the historical and cultural parks have attractions that are buildings or monuments. Rangers at these parks may spend all of their time indoors. Park rangers, like many other federal employees, should be prepared to move several times during their careers.

Park rangers are Civil Service employees, and employment is arranged through the Office of Personnel Management. Rangers usually begin at the federal level designated GS-5, although some positions are filled at GS-7. From the entry level, rangers may move through the ranks to become district rangers, park managers, and staff specialists in interpretation resource management or a related area.

To qualify for a GS-5 position, an applicant must have a college degree or three years' experience in park or conservation work. A combination of experience and college studies may be substituted for these requirements.

Technical Support Positions

The National Park Service hires a number of aides and technicians. These technicians have basic forest management duties. They guard against forest fires and watch for signs of destructive insects or disease. They lay out and help construct trails and fire roads. They mark trees for removal or pruning and conduct replanting operations. During the summer and fire season, the NPS hires additional forestry technicians on a temporary basis.

Park aides or technicians are more numerous than park rangers. Park technicians help the park rangers with all of their assignments. They work on plant and insect control projects, soil conservation teams, and fire-fighting crews. In historic and archaeological areas, technicians carry out plans to preserve and restore buildings and sites. They operate campgrounds, including such tasks as assigning sites, replenishing firewood, performing safety inspections, and providing information to visitors. They may also conduct guided tours, direct traffic, join road patrols, and operate radio-dispatch stations.

All aide and technician positions are Civil Service jobs, and employment is through the Office of Personnel Management. Applicants should check with the National Park Service Regional Office in the area in which they want to work. Park aides can start at federal levels designated GS-1, GS-2, or GS-3. Most start at GS-2. A high school diploma or some related experience is required. Technicians usually begin at GS-4 and must have completed two years of appropriate experience or two years of college.

Outlook for Employment

Like all federal agencies, the NPS budget is at the mercy of Congress. In recent years federal budgets have been tight, and only a minimum number of new employees have been added each year. More than three thousand park rangers are employed throughout the National Park Service. At peak employment during the summer season, the number of rangers stays about the same, but part-timers swell the NPS ranks. Forestry students should keep the NPS in mind when looking for summer jobs. Graduates of forest technician schools may find good jobs in the NPS. Professional foresters—especially those with a major in outdoor recreation, communication, or interpretation—can find satisfactory work as park rangers. In either case, expect the competition to be keen for the few jobs available.

United States Fish and Wildlife Service

The Fish and Wildlife Service (FWS) dates back to 1871, when Congress established the U.S. Fish Commission to respond to the decrease of the nation's food fishes. It is now the principal agency through which the federal government carries out its wildlife programs. It is responsible for managing the nation's wild birds and mammals as well as fish for the enjoyment of all people and for the betterment of the environment. The FWS is responsible for migratory birds, endangered species, certain marine mammals, and freshwater and anadromous fish. The Fish and Wildlife Service also fights the illegal importation and sale of protected wildlife from other countries.

The FWS has seven regional offices, including one in Alaska, and nearly seven hundred field units and other installations. There are almost five hundred national wildlife refuges totaling more than eighty-eight million acres. The FWS maintains more than one hundred wildlife research stations, including laboratories at thirty-eight different universities.

Since the management of wildlife requires the scientific manipulation of land and plants, there are a variety of resource-specialist jobs available in the FWS. This is obviously a place for wildlife biologists. Foresters with wildlife specialties are often prime candidates for employment. Sometimes forest managers or range specialists are hired. Natural resource management aide and technician jobs are sometimes available at the FWS.

The Bureau of Indian Affairs

This federal agency was created as part of the old War Department back in 1824. In 1949, responsibility for the Bureau of Indian Affairs (BIA) was transferred to the Department of the Interior.

Through treaties and other agreements, the federal government has formal relationships with and responsibilities toward the 562 tribal nations in the United States. For example, the government is to protect and enhance Native American lands and natural resources. Through the BIA, the government provides technical assistance in forestry, water rights, range management, irrigation, soil conservation, and mineral management. In addition to lending this expertise, the BIA administers more than forty-three million acres of tribally owned land, more than ten million acres of privately owned land, and almost half a million acres of federal land in trust status.

The BIA maintains a small staff of technicians and professional resource managers, including foresters. Even though forestry jobs are not plentiful here, a forestry graduate who wants to work with a small agency and have broad responsibilities should not overlook the BIA. Job applicants who can prove that they are at least one-fourth Native American are given preference.

Some tribes have extensive timber holdings and sophisticated management systems. They sometimes employ professional foresters to work directly for the tribe. This is a little-known opportunity for nongovernment, nonindustry professional employment.

Other Federal Bureaus and Departments

The agencies just discussed offer the best prospects for recently graduated foresters. But a survey of the many federal bureaucracies reveals a surprising number with foresters on their staffs. Students working toward forestry degrees and recent forestry-school graduates should know that jobs could come from unexpected quarters in the federal government.

The *Army Corps of Engineers* is a division of the U.S. Army. The Corps builds dams, bridges, canals, and water reservoirs. It manages more than five million acres of public land. Much of this property is prime parkland, riverside woods, and forest. The Corps hires foresters with engineering, management, recreation, and wildlife specialties.

Water resource development is the mission of the *Bureau of Reclamation*. Its western-state water projects are designed to provide wildlife habitat and recreation as well as water. Foresters are needed to manage the timber and other resources associated with these facilities.

The *Bureau of Mines* and the *Office of Surface Mining* are interested in getting coal and minerals out of the ground. They also are increasingly concerned about repairing the earth after the miners' bulldozers are finished. That is where foresters and range specialists are needed. As more and more land is ripped up to meet this country's energy and mineral demands, more foresters will be needed to help put the land back together.

The *Department of Transportation* sometimes hires foresters to help design highways through wooded areas, to landscape roadsides, and to help reduce the environmental impact of highways.

The *Environmental Protection Agency* has a few foresters on its staff. They are involved in the research and assessment of pollution's effects on forests.

The *Tennessee Valley Authority (TVA)* is concerned with the development of the natural resources in the seven states intersected by the Tennessee River. Its dam and reservoir system regulates the river and major tributaries for flood control, navigation, and power production. The TVA also is active in fish and wildlife development, outdoor recreation, energy research, and reservoir ecology. Multiple-use-minded foresters sometimes are hired to administer these projects and to conduct research.

The TVA has a Division of Forest Relations, which is divided into three branches. The Forest Development Branch is involved in forest development and watershed protection. This branch operates two large forest nurseries. The Forestry Investigations Branch surveys and assesses timber resources. It also conducts research in forest management, economics, and utilization. The Fish and Game Branch is responsible for managing wildlife resources.

There are many other agencies that sometimes hire foresters. The Office of Personnel Management handles most of these opportunities. Information about working for any of these agencies can be found at a Federal Job Information Center. Ambitious job seekers may discover unadvertised possibilities by writing to the agency's headquarters in Washington, D.C., and by visiting the nearest regional or field office.

State Forestry

State foresters are found in forty-nine states, eight U.S. territories, and the District of Columbia. The agency they work for might be part of the parks division, the conservation department, or some other natural resources organization. The structure and titles of forestry agencies are different in each state. The leading agencies are listed in Appendix E.

The size and importance of the state forestry agency is often relative to the importance of the forestry industry in the state. As you would expect, forestry is more important to Georgia and Oregon than it is to Kansas or South Dakota. Many states own large tracts of forestland that must be managed. Every state has some sort of forestry program that encourages landowners to plant trees and take proper care of them.

Because a state forest belongs to the public as a whole, it is managed for multiple use. The sale of timber may be a significant source of revenue for the state. Watersheds are always important, as every state needs reliable supplies of clean, fresh water. Many state forestlands are heavily used as parks. Recreational assets are given prime consideration. The forest also is managed as a fish and wildlife habitat.

The state forester (sometimes called a service forester) carries out the management policies of the lead forestry agency. Like any forester in the public sector, he or she must balance the many current uses of forestland and plan for the future. The state forest must be protected from fire, insects, and disease. Education of the public on the value of the forest is also a big part of the job.

Most states are divided into regions, much like the ranger districts in national forests. A senior professional forester administers each region and usually has a staff or other professionals and technicians. Some states operate tree nurseries and conduct reforestation operations on both public and private land. Most states have fire-fighting outfits that cooperate with the federal fire suppression crews. Some states have forest research efforts, which are often connected to federal facilities or attached to a university.

The state forestry organizations are an important part of this country's effort to make the best use of our timber-producing lands. They will be even more important in the future. This is because about 58 percent of the nation's commercial forestland belongs to private individuals. The publicly owned timberlands, such as national forests, are managed by government foresters. The industry-owned timberlands are intensively managed by company foresters. But who is managing that 58 percent of the nation's wood owned by private individuals? Nobody. Almost everyone in the business agrees that these undermanaged private lands hold the key to this country's future forest productivity.

The reasons why landowners do not manage their property for wood production vary, but leading foresters are convinced that if more landowners only realized what an asset their woodlots could be, more would emphasize wood production. Farmers may understand the economics and administration of corn, soybeans, or dairy cows, but few landowners know much about timber production.

In recent years, increased attention has been lavished on the private woodlot owner. The Forest Service is expanding its consulting and advisory service. Extension Service foresters are also available to landowners, but it is the state forester who is expected to take the lead in this effort.

State foresters are able to meet and consult with landowners to discuss how they can best harvest, replant, and maintain their woodlots. The forester can walk through a landowner's woods and recommend approaches for conservation logging or offer advice on which acreage is best suited for timber production. The landowner will have questions about government subsidies, tax benefits, and market potentials. State foresters also advise owners and operators of small lumber mills and other wood-processing plants. They can show the woodcutter how to operate more efficiently. Helping the wood processor or woodlot owner produces more benefits to the local economy. This assistance also increases the private sector's contribution to the nation's wood needs.

State foresters coordinate the various landowner-education programs of industry, federal government, and local government. To bring these ideas and efforts together, the National Association of State Foresters was formed to actively promote programs that offer technical aid to local landowners. They are also trying to draw public attention to the need for a vigorous state forestry organization.

The Outlook for State Forestry

More than five thousand foresters are presently employed by the states. State forestry programs are expected to remain strong and to expand. The amount of expansion depends, year by year, on government budgets. In recent years, there has been a trend to turn federal programs over to the states. That trend will benefit state foresters, possibly at the expense of federal forestry programs. At the same time, state budgets are being stretched to the breaking point. So the number of new opportunities in state forestry will fluctuate with the economy and the presidential administration. To find out more about the forestry program in your state, write to your state forester. The addresses for state foresters are located in Appendix E.

County Forestry

Local communities have stepped up their efforts to preserve, maintain, and responsibly utilize the forests in their jurisdictions. As a result, new opportunities exist at the county and township levels for professionally trained foresters. The forests they are called upon to manage may include county parks, watersheds, school grounds, recreation areas, and even the trees that line the streets. There are several thousand community forests in the country, totaling about eight million acres.

Some counties have sophisticated conservation and land-use programs under the supervision of a county forester. These often include cooperative ventures with private landowners. The forester oversees the cutting of timber from county property and replanting efforts. He or she also helps plan future public facilities such as playgrounds or highways.

A county or township area that has control over its own watershed often hires a forester to see that the area is properly managed. A watershed forest can funnel quality rainwater to a reservoir while sustaining logging and recreation. To serve these many purposes, a watershed, like any forest, must be properly managed.

Urban Forestry

As urban populations have grown, the wooded areas in cities have come under great pressure. Yet it is the trees as well as the architecture that gives a city its character. With this in mind, the city of Chicago, for example, undertook a massive tree planting in recent years, inspired by the gracious charm of Parisian parks. A wooded park is often the most popular public place in town. For example, life in cosmopolitan New York and romantic San Francisco revolves around the expansive central parks in each city. Trees make property more attractive, and thus more valuable. City streets or suburban avenues would seem cold and barren without trees, which bring a refreshing touch of nature to the canyons of concrete, steel, and glass.

Trees serve more than an aesthetic purpose. They save energy, providing shade in the summer and windbreaks in the winter. Their roots hold soil against erosion. Trees actually breathe carbon dioxide in and oxygen out, thus reducing air pollution. They muffle the noise of traffic and construction, and the soft rustling of their leaves in the wind soothes us.

Communities are always proud of their trees. Sometimes the trees are even tourist attractions. When spring releases the cherry blossoms in Washington, D.C., people come from miles around to enjoy this sweet-scented and short-lived phenomenon. The flame trees in some small Oklahoma towns have the same effect. New England villages and towns celebrate and benefit economically from the spectacular displays of fall foliage.

Communities will defend their trees. One highway project threatened a row of ancient maples in an Oregon college town. The trees were to be cut down so the road could be widened. Nearby residents chained themselves to the trees to protect them from the saws.

Many urban governments hire foresters to look after their cities' trees. The job these foresters do is quite different from the duties of a typical woodlands manager. Urban foresters often work with small groves or even individual trees. They also work constantly with people.

Trees are sometimes the oldest residents in a neighborhood. They are often landmarks and gathering places. The forester strives to keep these celebrated old-timers alive and healthy. It also is the urban forester's responsibility to call for a tree's removal when it begins dropping dead limbs, threatening the safety of passersby, or when it has grown to the wrong shape or size and can no longer be "fixed" by pruning.

Big-city foresters spend a lot of time in the city-planning department. They have the important function of choosing the right tree for the right place. Cities are always growing and building. Foresters try to direct new construction to make sure that existing trees are not destroyed unnecessarily. They can also recommend trees for planting around new construction, as well as offer advice on their care.

Trees in the city face unusual hazards. Air pollution, poor soil, climbing kids, and power lines are not problems for the typical tree in the forest. An urban forester knows which trees can tolerate the demands of city life.

Urban foresters also provide community education regarding the local trees. Home owners and caretakers of institutional grounds need to know which trees to plant and how to care for them.

Urban forestry is becoming a recognized science. City council members are usually surprised to hear an urban forester explain how every tree in the city can be cared for within one comprehensive management plan. They are amazed to find out that city trees can sometimes be harvested and the wood sold, and that the area can then be improved by replanting. New hybrids and introduced species can be superior in every way to the trees they replace. The city council listens with interest as the forester shows how all the trees in the city are really part of one big forest—a forest that can be managed to improve the quality of life environmentally, socially, and economically.

The Outlook for Urban Forestry

Urban forestry is an unusual and challenging way to put a forestry degree to work. The opportunities are increasing dramatically, though many cities are having difficulty affording even such essential services as fire protection, police, and trash collection. Nevertheless, forestry programs are receiving greater and greater attention. Although cities rarely hire more than a few foresters, the number of cities doing so is on the rise.

Both the Forest Service and the American Forestry Association (AFA) actively promote urban forestry. The AFA has created a software tool, CITYgreen, to help people understand the value of trees in their environment. It also sponsors Global ReLeaf, an education and action program that promotes planting trees in the United States and around the world. The trees are planted to cool and beautify the neighborhoods, help with restoration of the ecosystem, and slow global warming. Special campaigns might include planting trees in war-ravaged cities such as Sarajevo, or encouraging and assisting people to plant native trees in forests damaged by wildfires.

Forestry experts expect urban forestry to be a wave of the future. They are gratified to see more and more cities recognizing the value of managing city trees. And, since urban development is consuming millions of acres annually, the urban forester's land base is growing. Meanwhile, the traditional forester's timberland is steadily shrinking. The optimism for the future of urban forestry may well be justified.

Forestry Education

As a professional forester, one must have a strong body of knowledge regarding many aspects of forestry. Equally important is having the ability to communicate this information. As a forester, you will find that you are also a part-time educator. Industry foresters must explain silvicultural techniques to subordinates. That is education. Explaining your company's plans to the residents of the area is education, too. Government foresters are particularly called upon to deal with the public, answering questions and promoting conservation. The classroom might be a campground, a civic auditorium, or the woodland itself. Government foresters are often invited to talk in schools, from elementary school classrooms to graduate seminars.

Other foresters, such as extension agents or interpretive specialists, teach. And some of these teachers specialize in forestry. There are more than a hundred universities and technical schools in the United States with forestry programs. These schools all have staffs of teachers specializing in one or more aspects of forestry. The size of these teaching staffs varies from a half dozen up to more than fifty, depending on the size of the institution.

The four general categories of forestry education are universities, public or technical schools, the Extension Service, and interpretive forestry for government or industry. To be effective in any category requires that the teacher be intimately familiar with the subject. Teachers also must enjoy working with people.

University and College Teaching

Some forestry students are so inspired by the process of attaining their undergraduate degree that they decide to pursue a career in academia. Although a doctorate is not absolutely necessary to teach at universities or colleges, opportunities to teach without at least an M.S. degree are increasingly rare, and a doctorate is usually required, except at two-year schools. Several years of "on the ground" experience also are desirable, though not essential.

Forestry professors usually specialize in one or two related subjects. Some teach general forestry courses such as elements of forestry, forest management techniques, or forest ecology. These courses are usually for the first- or second-year student. Other professors teach only highly technical courses, such as principles of forest modeling or forest biometrics.

Many forestry courses require field trips to the woods or to industrial sites. Most schools with active forestry programs have a representative forest nearby. Some of these holdings are quite extensive. The forest is used as an outdoor classroom; the students may actually manage it for profit and multiple use.

College teachers are part of an academic community. In addition to teaching, they are expected to contribute to the administration of the school and the development of programs. They usually have the opportunity to engage in individual research. Some university professors direct the research projects of graduate students. The professor also acts as a counselor, tutor, and role model. Professors at land-grant (agricultural) colleges also may serve as county or state extension agents.

A college teacher often begins as a part-time laboratory or teaching assistant while still completing his or her degree work. The first full-time position might be as an instructor. From there, one generally moves up the ladder to assistant professor, then associate professor, and finally full professor. A teacher with exceptional administrative skills may eventually be chosen as head of the department or dean of the college.

College teachers have traditionally earned less than their counterparts in industry. This difference in salary is offset somewhat by the professor's ability to choose working hours, conduct research, and take sabbatical leave. Academics are sometimes also able to supplement their income by serving as consultants to private and public organizations. Perhaps the greatest appeal of a career in education, however, is the chance to contribute to the growing body of forestry knowledge.

Teaching in Public or Technical Schools

As more and more people become interested in and committed to conservation and environmental protection, a growing number of high schools and community colleges have begun to offer courses in ecology. These courses usually cover basic ecological relationships and the conservation ethic. Forestry might be only a small part of the instruction.

Forestry is a discipline that requires looking at many environmental factors over time. It is a study of large ecosystems—the "big picture." Foresters see the lessons of ecology in action, so many foresters are well suited to teach basic conservation in the public schools.

The demand for technicians in the natural resource management field is high. Community colleges, vocational training centers, and ranger schools have found that their forestry programs are quite popular. These schools concentrate more on technique and less on theory than university programs. Their courses are largely hands-on courses where students learn by doing, so the instructors are valued for their experience as well as their academic credentials.

The Extension Service

Extension agents serve as a liaison between research scientists and the public. They are funded by cooperative agreements between the U.S. Department of Agriculture, universities, states, and counties. The extension agent's pupils are usually adults: home owners, farmers, and businesspeople.

The agent works out of the university or a county office. The agent who is a forestry specialist monitors the findings of university scientists and Forest Service researchers. This information is then passed on to local timberland owners. The agent writes brochures, gives talks, and answers letters and phone calls. He or she is frequently asked to visit a woodlot or nursery and give advice on site. This is one-on-one education. The extension program is valuable and highly praised.

Interpretive Forestry

One of the responsibilities of government foresters managing timberland is to explain the rationale of forestry programs to the public. The interpretive forester is a specialist who provides information to help people better understand and thus enjoy the forest. He or she leads the public to an understanding of forest ecology. In other words, the interpretive specialist is a teacher whose classroom is the outdoors.

Although the timber industry has no legal responsibility to educate the public in forestry matters, most companies have found it to be good business to explain their timbering activities to the local people. For some companies, it's not just good public relations but also a community obligation. The timber business involves forests, and forests affect people. Even local citizens who do not own timberland or work for the logging company still have strong feelings about what happens in the woods. They have a sense of ownership and connection to the woods they see every day, even though they might not literally own them. The company forest may be the community's watershed or home for the local wildlife. It is certainly part of the scenery. The company's interpretive forester educates the public on the timber business and forest conservation and gives the public a way to talk with the company. In this exchange of information, ideas, and opinions, everyone benefits.

Private Industry

Privately owned industry is often the best place to look for a job as a professional forester. It was not always so. In the 1930s, there were only a few hundred foresters working outside of the government. Even now, the majority of foresters work for federal and local governments. However, great opportunities exist for the 25 percent of foresters who follow the path into private industry. They work for logging and lumber companies, sawmills, research and testing services, and other forestry-related businesses.

Companies that own timberlands are forced, by economics, to manage those forested acres intensively. That means hiring more foresters than in the past. Just a few years ago, a single forester might have been responsible for 100,000 acres of company lands. Now a manager may have 25,000 acres or less. In Europe, where forests are small and closely managed, a forester may work with fewer than 5,000 acres.

Forest products companies own huge blocks of timberland. They control, contract, lease, or cooperatively manage millions of acres owned by individuals. The forester's job is to make these woodlands produce the raw materials needed for the company mill. Some companies specialize and produce just one product, such as plywood. Most of the larger landowning companies are diversified, making a variety of wood products. It is useful to divide wood products companies into two general categories—those that produce mostly boards ("dimensional stock") and those that are mainly in the paper business.

Paper manufacturers are necessarily large companies. A paper mill is a giant, complex facility. It demands a tremendous capital investment. Once it is built, it cannot easily be moved, so paper mills need a steady supply of wood coming in the gate. Foresters are employed to see that this flow of wood is not interrupted. They also must ensure that the wood supply is of high quality and obtained at a fair price.

Duties of the Company Forester

Foresters working for a paper or timber company draw management plans for the company woodlands. They determine which areas should be logged and when. They mark and scale (measure) timber. Foresters also supervise the logging itself. They lay out the roads necessary to get logs out of the woods. They examine the soil, inventory wildlife populations, and make sure that streams are protected from damage. After an area has been logged, foresters oversee the cleanup and replanting. Then they ensure that the new seedlings have every advantage for quick and healthy growth.

Some forest products companies own and operate tree nurseries and grow their own seedlings. A forester usually supervises these tree farms. The millions of young trees are used for reforestation projects on company lands. Most companies encourage private landowners in the region to plant trees, too. The company will make seedlings and planting equipment available at low or no cost.

A company forester may spend a lot of time working with individual landowners in the area. If the landowner agrees to sell timber to the company, the forester will draw up a management plan for the property. When the trees are mature, the area will be carefully logged. Then the forester oversees replanting efforts on the property. The goal is to have that same woodlot produce more marketable timber in the future.

The forester also watches woodlots bordering company property for signs of fire danger or insect infestation. Some companies spend millions of dollars every year guarding their timberlands from fire and insects.

Forest protection may be the company forester's first priority. The companies train and outfit fire-fighting crews, and during fire season, they maintain lookout towers and fire spotters in airplanes. They keep fire suppression equipment on the line and ready to go. If fire does break out, everybody goes to work. From the temporary laborer to the senior forester, everyone has a job to do. Some foresters are specially trained in fire suppression. They direct the attack. When it is over, the company foresters assess the damage and supervise salvage operations. Then they organize the replanting program.

Insects are a constant threat. There are a number of insects that can cause millions of dollars in damage in just a short time. The forester must always be alert for early warning signs of insect invasions. Some infestations can be controlled by management techniques such as selective cutting or spraying, but they have to be detected early. Once the insects take over an area, the forester's only recourse might be spraying pesticides. Since this is an environmentally delicate operation, the forester must supervise it carefully.

Foresters working for papermakers may spend much of their time contracting with lumber mills for by-products such as sawdust, chips, and mill ends. These wood scraps are unusable to the lumber cutters, but they can be made into pulp for paper. The paper company forester is always looking for new supplies of wood.

Wood technologists work in private industry seeking new uses for wood. They develop new wood products, new finishes, better adhesives, and longer lasting preservatives. In the paper or lumber mill, they look for opportunities to use different kinds of wood or wood by-products and ways to use wood more efficiently.

Wood technologists maintain quality control. They see that lumber or other wood products are properly seasoned and conditioned. They check the quality of raw materials and workmanship. In products factories, they ensure that the wood used is right for the job. In a furniture factory, for example, they check the moisture content, the way the wood is cut and glued, and the quality of the finish.

The industry also employs technologists in sales positions. Since the technologist knows the product inside and out, he or she is the best person to market it. The technician can also suggest ways to improve or adapt the product to make it more marketable.

The Outlook for Private Industry Employment

Industry employs the full range of foresters, technicians, technologists, and laborers. Because of budget cuts in government programs and restrictions on logging on public lands, many of the best opportunities for forestry school graduates will be found in industry. In the southeastern states of the United States, foresters will be particularly required to work with private landowners, helping them with forest management plans that increase production while also protecting the environment. Another area of growth is for research and testing firms, where foresters are needed to help prepare the environmental impact statements required by government regulations.

Self-Employment

Want to work for a boss who is competent, admirable, fair, generous, and progressive? Many people do enter the workforce and find such supportive and inspiring supervisors. Others say that if you want to find a boss with all these wonderful qualities, you need to be that boss yourself. Luckily, forestry is one of those fields in which you can be self-employed.

With a forestry degree, lots of ambition, sales skills, some business abilities, confidence, and a little luck, you can make it as a self-employed forester. Going on one's own also requires discipline and motivation, because it is not the easiest or most secure way to make a living. There are, however, so many different ways to work as a self-employed forester.

Consulting Forester

The most common form of self-employment in this field is consulting work. Consultants sell services and advice. The key to success is finding a service or body of knowledge that is not readily available elsewhere.

Services can be marketed to landowners who lack the personnel or equipment to do the work themselves. The consultant could be hired as part of a crew or to work alone. Salable services include such activities as:

- Cruising timber stands and marking trees to be harvested
- Scaling (measuring the volume of) harvested logs
- Pruning and thinning trees to encourage rapid growth

- Mapping soil or tree types
- Designing roads or logging facilities
- Surveying and laying out roads or harvest areas

The most valuable attributes of a professional forester are knowledge and experience. These can be marketed to timberland owners in need of some type of forest management. Since many small timber owners cannot afford to keep a forester on staff, there often is a need for a consultant. General consulting services that a forester can sell include:

- Making sure that the right trees are cut (and only the right trees)
- Helping a woodlot owner determine when to harvest and how much money to expect
- Determining the cost of damage done by fire, wind, or disease and advising the owner on how to salvage the timber
- Assisting the timber owner in drawing up a management plan
- Ensuring that the land will be protected from undue environmental damage during logging operations
- Advising a forest products company on locating a mill or on running a mill efficiently
- Preparing an environmental impact statement
- Performing administrative services such as negotiating timber sales, overseeing the harvest, supervising forest management activities, and administering a woodlot for an absentee owner

Much general advice can be obtained from state or federal foresters at little or no cost to the landowner, so consultants often develop a special kind of expertise that is not usually available from government sources. Such specialties might include authoritative knowledge of a particular kind of tree disease or insect, certain logging methods, uncommon machinery, reforestation of unusual sites, or ways to improve wildlife habitat.

Government or private industry sometimes hires consultants to conduct research. A Ph.D. and some research experience usually are required to land such a contract. Research contracts given to self-employed consultants may be for any kind of data gathering. The consultant may be asked to investigate a particular market, an economic situation, a problem with tree-girdling porcupines, or to determine why a stand of hemlock is not meeting growth expectations.

Outlook for Self-Employment as a Consultant

The outlook for consulting work is considered good to excellent. Recently graduated foresters should be cautioned, however. Timber owners are sometimes reluctant to hire a consultant fresh out of school with no track record. Also, the consultant market is sometimes flooded with foresters newly retired from government or private industry. As an alternative to being a self-employed consultant, you might consider joining a consulting conglomerate. These companies employ a team of consultants with a variety of specialized skills. These firms offer professional opportunities for foresters regardless of experience.

Sawmill Operator

Another way to work for yourself is to run your own sawmill. As a graduate of a technical school or of a wood products curriculum, you would have the knowledge to operate a successful timbering operation.

There have been tremendous technological advances in the sawmill industry. In some modern mills, the incoming logs are scanned by video cameras. Computers determine how the log can be cut most efficiently. Razor-thin saw blades, controlled by robots, go to work. There is almost no waste.

Specialty sawmills cut particular products such as veneer for plywood or furniture, barrel staves, and shingles or shakes. Other mills cut only certain kinds of wood. The lumber mill makes logs into boards for construction. Even these mills may specialize in boards of a certain size or type.

There are portable sawmills that can be hauled from site to site. Some mills are taken right into the woods. The operator is paid for the number of board feet that are cut. Other mills are stationary, and the operator buys logs and hauls them to the mill. Some mills work in conjunction with a lumber retail yard.

The enterprising forester can get involved with milling at any level. Some foresters operate mills and use their knowledge of timber harvesting and marketing methods to ensure that they can get a steady supply of logs at a reasonable price. They are involved in the whole process, from standing tree to marketed product. Others concentrate on one phase. The potential for self-employment in sawmill operations fluctuates. When the housing market is strong, demand for timber is high. But a slump can be disastrous for the small, undiversified mill operator. Some specialty products command a steadier, if more limited, market.

Logging Contractor

Logging companies often need the services of a professional forester. A forester must survey and evaluate the stand to be cut. Boundaries must be marked, seed trees identified, locations for logging facilities mapped, roads designed, cut logs scaled, and contracts negotiated.

Logging is hard and dangerous work, best left to loggers. It is not the kind of thing you learn to do on a Saturday afternoon. But if you are both an experienced logger and forester, you can roll the two jobs into an efficient and profitable business.

One drawback to self-employed logging is the incredible cost of the necessary machinery and equipment. There certainly are risks to becoming a logging contractor, but the forester should have a better chance than most of making it.

Christmas Tree Farmer

It used to be that most Christmas trees were cut from public and private forests. The current demand for evergreen trees is now increasingly met by plantation Christmas tree enterprises.

A tree grower who specializes in Christmas trees can produce trees that are just the right size and the perfect, conical shape to decorate a home or business. The trees are spaced when planted so that they do not crowd each other. By careful spacing, the farmer can get three thousand trees on a single acre. The trees can be pruned every year or two to force them into the right shape. Shearing also causes the foliage to thicken. Favorite species include Douglas fir, grand fir, noble fir, black spruce, white spruce, and red cedar.

Seedlings that are two or three years old can be purchased from a nursery. Then it takes another seven to ten years before harvesting can begin. The trees are cut as early as October and shipped, sometimes thousands of miles, to Yuletide markets. Some tree farmers allow families to visit the plantation and cut their own trees.

Christmas tree growing is a good sideline business for anyone interested in working with trees. A forestry education should be useful to the tree farmer. The production and management techniques are much the same for both the woodlands manager and the Christmas tree farmer.

Other Opportunities for the Self-Employed

There are many other ways to find self-employment in forestry. In fact, only your imagination and resourcefulness limit the opportunities. We will confine this discussion to those jobs that are primarily silvicultural, using forest management techniques, and/or those jobs that make good use of a forestry education. In addition to those vocations already mentioned, some self-employment possibilities are described below:

Wood Products Marketing

Not many people can start their own paper mill or timber company, but there are lots of ways for the small businessperson to get involved in the harvesting and marketing of forest resources. Small operators can best handle some secondary forest products.

Here is a list of such products:

- Fireplace wood
- Specialized woods for art objects and furniture
- Gourmet wood charcoal
- Maple syrup
- Poles, pilings, and split rails
- Fence posts
- Shingles
- Tree seed
- Pharmaceuticals

Landscaping

There is a demand for landscape architects who plan and supervise beautification projects for government, business, and home owners. Landscape gardeners do the actual work, and they, too, know much about the physiology of trees.

Tree Surgeon

The "tree doctor" is needed in urban and suburban areas to prune, fertilize, and spray ornamental trees. Like landscaping, this work is physically demanding and requires a familiarity with certain tools and machinery.

Tree Nursery Operator

There are several different kinds of tree nurseries. Some grow seedlings of forest species for reforestation projects. Some raise young fruit or ornamental trees. Others supply landscapers with a wide variety of tree species. Many nurseries are run by government agencies to supply public forestry agencies. There also is room in the market for the private operator.

Associations and Organizations

A number of trade and professional organizations have foresters on staff. Many of these groups represent some faction of the forest products industry. Others represent the professional forester.

There are state, regional, and national trade associations and organizations. The members may be pulp and paper companies, wood products companies, tree farmers, or logging contractors. The members have organized to promote an awareness and understanding of the business. The groups conduct research, public education, or lobbying. Many regularly publish journals, brochures, or other information. The American Forest and Paper Association is one example of a trade association.

Practicing foresters often belong to a group like the Society of American Foresters, the Canadian Institute of Forestry, or the American Forestry Association. These organizations strive to promote professionalism, professional ethics, and coordination among members. They also promote the concepts of forest management and resource conservation, and they educate the public about forestry. Through conferences and professional journals, the associations provide an exchange of ideas and new information among foresters.

These trade and professional associations hire a small number of exceptional foresters. It is a feather in the cap of a forester to be hired by a professional or trade association. These positions are generally not available to recent graduates. A forester usually must have a good record of field experience before becoming a candidate for association work. The work usually is in administration, education, public information, or, less frequently, research consulting.

Citizens' conservation and lobbying organizations sometimes hire technical staffs to analyze current conservation issues, to define policy, and to educate the public. The National Audubon Society, National Wildlife Federation, Sierra Club, and the Wilderness Society are examples of this kind of organization.

These groups are not-for-profit, citizen-action organizations supported largely by donations. Foresters working for nonprofit organizations generally do not earn as much as those in the private sector. But for those who are interested in advocacy and politics, it is an excellent way to become acquainted with the political implications of resource management decisions.

Interest in such public service positions is high. Competition for the very limited number of jobs is keen. If you have strong communications skills, a willingness to work long hours, and a dedication to promoting conservation, do not overlook the job possibilities in associations and organizations.

7

Preparing for a Forestry Career

As you have seen, forestry is a profession with many disciplinary paths. These can lead to the woodlot, the boardroom, or the classroom. Or they can take you in a multitude of other directions, wherever you chart your own course.

There is no one guaranteed or required way to prepare for a career in forestry. However, there are activities and experiences that can serve as first steps to becoming a professional forester. As a child, participation in scouting and mountaineering groups develops one's resourcefulness, independence, and leadership skills. Hunting and fishing build outdoor skills and foster an appreciation of nature. Foresters have traditionally been grown-up versions of children who loved camping, fishing, hunting, and hiking in the woods. Although it may seem ridiculous to begin planning a career too early, it is probably worse to choose an occupation that you do not find interesting. The best career for you might just be an extension of those things that have always aroused your sense of curiosity and adventure.

Most young people begin the process of zeroing in on their life's work during their last two years of high school. Some tough decisions must be made during these years: Do I go on to a university? A community college? A vocational center? Join the armed forces? Go right to work? If I go to college, which one? What should I study? The answers to these questions, obviously, will have a profound influence on your adult life.

If you decide to seek a college degree in forestry, you will study a wide variety of subjects. A typical curriculum includes science courses such as silviculture, wood science, chemistry, statistics, genetics, meteorology, plant disease, entomology (study of insects), dendrology (study of trees), fisheries science, soils, surveying, wildlife ecology, park management, physics, and botany.

But a forestry curriculum is not all science. To really prepare for a career in forestry, you will also enroll in liberal arts courses that will provide the foundation for communicating with colleagues and the public and will enable you to better understand the social and political worlds. Courses that may prove very helpful to you in the future might include psychology, technical writing, photography, history, economics, speech, or world literature.

Then there is the decision over whether you want to focus on a particular specialty area within forestry. This often is done after the first or second year of school. Program options include forest management, forest engineering, forest products, outdoor recreation management, and range management.

Forestry schools provide a good general education. The successful forester has to know a little bit about a lot of things. If, for some reason, as a forestry school graduate, you decide not to become a forester, you'll be well prepared for many other jobs. Some graduates make good use of their education by going into business, farming, construction, teaching, or other fields. There are also enticing opportunities abroad for natural resource managers.

Professional Forester

If you have decided to become a professional forester, you should make certain that while you are still in high school you are taking the courses required for college admittance. These requirements vary greatly but usually include a full range of science, math, languages (including English), and social studies. College-bound students should have a strong academic record, including both good grades and adequate scores on standardized tests.

Some high school students worry about which courses to take to prepare for their chosen college curriculum. Students preparing for the professional forestry program have it easy. Forestry school deans always advise their future students to get a well-rounded education. Concentrate on science, math, and communications—the basics. Fortunately, for the future forester, these are the same recommendations made by the dean of any school in the university. In other words, preparation for forestry will also ready you for most other fields.

The student also should make provisions for meeting the high cost of a university education. Gaining admission to a university is rightly considered an honor, but there is no reason why any determined student cannot go to college. Forestry students can work at part-time jobs to help pay college expenses. Summer jobs in the forestry field not only help meet expenses, they also provide valuable work experience. For the exceptional student, there are scholarships, fellowships, assistantships, loans, and grants available.

If you are a high school student considering a career as a professional forester, you will have to reckon with two very real challenges—grades and finances. Your high school guidance counselor can give you good advice on how to proceed.

What if you have already enrolled in college in some department other than forestry? If you are an undergraduate who has just rekindled an old love for trees and the outdoors and you are thinking about switching to forestry, take heart! Regardless of what you have been studying, you can use that course work in the forestry department. Foresters who have had a couple of extra years of schooling in fields like business, economics, computer science, math, education, or journalism have a tremendous advantage. Forestry is so diverse that it can provide accommodating niches for almost any combination of scholastic backgrounds. Even studies such as foreign languages and international relations can be profitably combined with a forestry degree. Transfer students from any of the basic sciences—such as physics, biology, or chemistry—will, of course, find their credits good in forestry school.

If you are a graduate of some program other than forestry, you may find that pairing your degree with a forestry degree can win you that perfect job. Almost any combination of degrees can be put to work. An engineering or law certificate wedded to a forestry degree is an especially powerful combination.

There is no reason why you must follow a prescribed or traditional path in seeking a forestry education. The traditional path leads directly from high school to four years of college to work. Most variations of this route are advantageous. It is okay to switch majors. It is fine to drop out of school and work (in the forestry field) for a year. It is an acceptable practice to get a B.S. degree, work for a couple years, focus your interests, and then return to school for another degree. Do not feel bound by tradition, but do not let an interruption of your formal education deter you from your goal, either.

A professional forester never stops going to school. Forestry is a rapidly evolving science. The forester who wants to advance must stay abreast of new developments. You can accomplish this by reading professional journals such as *American Forests*, *Journal of Forestry*, and *Forest Science*. Foresters also attend professional conferences, seminars, and occasional refresher courses.

See Appendix D for a list of colleges and universities in the United States and Canada that have forestry programs.

Technician

The high school student who is more comfortable working with tools than with theories might consider a technical or vocational school. These institutions, sometimes called "ranger schools," have one-, two-, and three-year programs. The two-year course is the most common. Technical schools prepare students for paraprofessional forestry work. Technicians usually aid the professional forester. Their duties are discussed in detail in Chapter 5.

Recommended preparation for technical schools is essentially the same as for universities. Focus on the basics and take responsibility for getting a solid and comprehensive education. Vocational schools usually are more flexible and lenient about their entrance requirements. Students planning on becoming technicians can also take more shop and craft courses than university-bound students.

It is possible to transfer from most technical schools to a forestry program at a four-year college, but be forewarned: if you are considering this route to a forestry degree, make certain you know in advance how many credits you can transfer from the technical center to the college. Each school has its own policy.

Technical schools often are much less expensive than universities. Since the programs are shorter, the student can begin working sooner. See Appendix C for a list of forest technician schools in the United States and Canada.

Immediate Employment

Some high school students want to begin working immediately after graduation. Some intend to go to college later; others decide not to go at all. There are jobs in both government and private forestry for workers without special training. These usually are laborer or aide positions.

If you cannot offer an employer a degree or technical skills, your next best pitch is to demonstrate an ability and willingness to work. Determination and responsible behavior can get you a long way, but eventually your career progress will be blocked by a lack of academic or technical training. At that time you may want to go back to school.

Other students choose to enter military service after graduating from high school. For some this is an ideal course of action. The armed services may provide you with a skill that later can be employed in forestry. It will certainly give you an opportunity to mature and develop leadership capabilities. Veterans' benefits are another advantage of completing military service. Veterans are given financial assistance that helps defray the cost of a college education.

AmeriCorps

If you are ready to go to work but not quite ready for college, another option is AmeriCorps. Here, as with military service, you can acquire skills that will prove useful in a forestry career. If you were to join up with AmeriCorps, you would work with one of the national service programs throughout the country. These programs are focused on responding to needs in education, public safety, health, and the environment. AmeriCorps members engage in a wide range of community-oriented projects, such as building housing for the poor, cleaning parks and streams, or helping communities recover from disasters. Most AmeriCorps programs are open to high school graduates and provide at least a stipend, sometimes even housing. Following a year's full-time service, AmeriCorps members will also receive an education award of $4,725 to pay for college expenses.

The Peace Corps

University or technical center graduates have an attractive alternative to entering the job market—the Peace Corps. Resource management skills will always be urgently needed by developing countries. The Peace Corps recruits people with these and other skills for service both overseas and in the United States. It is an excellent opportunity to learn foreign languages and to experience other cultures. The Peace Corps makes everyone a teacher of skills and a student of culture. For students interested in interpretive or international forestry, the Peace Corps can be a valuable training ground.

Blazing Your Own Path

If you have committed yourself to a career in forestry, the path you take to get there is up to you. Regardless of your age or educational level, just about every option is open to you. Each option has its own advantages and disadvantages. Finding the right fit depends on knowing your own strengths and weaknesses—and what you are able to afford in terms of time and money.

Fortunately there are plenty of people around who are qualified, and willing, to give you advice. School guidance counselors are an obvious source of information and advice. If you are presently enrolled, they know you personally, and you can ask them to help you weigh the advantages of various courses of action. If you are not enrolled presently, you can request help from the school you are considering.

If you are thinking about attending a forestry school, gather information early. Write to the school and ask questions. Visit the forestry school and talk to students and faculty members.

The success and satisfaction you find in any career depends, to a great extent, on preparation. Since forestry is such a diverse field, not everyone has to be prepared in the same way. There is no standard mold for the forester. Determine what kind of forester you want to be, then set out to be the best.

8

Job Prospects

Maybe you are just now beginning to plan for a career in forestry. Soon, perhaps, you will be in your first year of forestry school. There is a lot to accomplish between now and then, but it's never too early to start dreaming about what your future career will be like. And it's natural as well to have some concerns about what your job prospects will be.

As it turns out, this is an interesting and challenging time to be in forestry. There will be wonderful opportunities, but it will be up to you to prepare yourself to stand out in a competitive field. According to the Bureau of Labor Statistics, employment of foresters will grow more slowly than the average for all occupations through 2010. The best opportunities for forestry graduates will be in state and local governments, as well as working for research and testing services. These organizations continue to seek bright and ambitious foresters to help them respond to the growing demand for environmental protection and responsible land management.

The federal government has long been a steady source of employment for foresters. However, changes in public policy can have an impact on the demand for foresters. The Bureau of Labor Statistics points out that federal land management agencies, such as the Forest Service, have moved their emphasis from timber programs to projects aimed at wildlife, recreation, and sustaining ecosystems. These programs still require the expertise of foresters, though perhaps not as prominently as in the past. On the other hand, many government positions in forestry will open up over the next ten years as the result of a wave of retirement among older foresters.

Some of the best forestry positions will be found in the southeastern part of the country, where much of the forested land is owned privately. Here and throughout the United States, private landowners will employ foresters to design and implement forest management plans. In addition, there will be sustained employment in private industry. Some foresters will be needed to work full-time as salaried employees for paper companies, sawmills, and pulp wood mills. Other businesses will draw on the skills and expertise of consulting foresters.

When the Bureau of Labor Statistics compiles data on foresters, it considers foresters and conservation scientists as a linked pair. Foresters manage the forested lands, whereas the conservation scientists (or range managers) are charged with managing and protecting the open rangelands. In 2000, foresters and conservation scientists held about twenty-nine thousand jobs. About 40 percent worked for the federal government, many of them for the Forest Service of the U.S. Department of Agriculture. State and local governments employed about 25 percent. The remainder worked for private industry, as private consultants, or as teachers.

Popularity of Forestry School

Forestry school has always been a favorite choice among high school graduates. Since the dawning of the environmental age in the late 1960s, all resource- or nature-related jobs have grown in popularity. That popularity faded during the 1980s, when young people were more interested in business, electronics, and hardware-related occupations. Today, forestry school has regained its popularity.

The Society of American Foresters (SAF) watches the job market closely. It has conducted a number of surveys to see how many graduates found work in the field. Its figures are not encouraging, as they tend to verify the notion that the forestry job market is always flooded with qualified applicants. The numbers seem to prove the forestry professor's droll statement: "Only half of you will find jobs." But these percentages are not much different from those of job-hunting success among other scientists and engineers. And the SAF survey showed that having advanced degrees considerably increases your chances of finding a job.

There are other things you can do to increase the odds. One of the most important is choosing the right specialization. Most foresters have a specialty these days, learned either in school or on the job. To be competitive in today's job market, you should pick up some specialized training before leaving school. After your second year at forestry school, you probably will be faced with a curriculum decision. Upper-division courses are usually more specialized than those taken in your first two years. The categories often are these: forest management, forest science, forest engineering, and wood science. The differences among these specializations are detailed in Chapter 5.

As the forest products industry becomes more competitive, those foresters who can help a company run more efficiently find their talents in demand. Engineers can spend, or save, a lot of company dollars, so good engineers are wanted. For this reason, foresters specializing in procurement (timber buying) and timber harvesting can usually find jobs.

Both industry and government employers want foresters who understand business and economics. Many recently graduated foresters have returned to school for master's degrees in business administration (M.B.A.). This combination can pave the way for a job in the timber company's corporate office. Another ticket to an administrative position is polished communication skills. Employers beg for foresters who can read and write well.

All businesses have been affected by the computer revolution. Forestry is no exception. Computers are being used in all areas from forest research to wood marketing. Forest products companies are hiring more and more computer specialists. A forester proficient in computer operation should have a good chance of finding a job.

Perhaps the most important change in the forestry job market is the new emphasis on wood science and technology. Forest products companies are moving the focus of attention from timber production to wood utilization. Instead of simply trying to grow more and more timber, industry is looking for ways to use wood better. There are jobs for wood science specialists who can improve manufacturing processes and develop innovative ways to make the most of wood. This new breed of forester is more likely to work in the mill than in the woods.

Today's forester may find this rule to be generally true: jobs are more available in the office or mill than in the woods. The most promising specializations seem to be engineering, economics and business, wood science and technology (utilization), and computer science. Graduates with specializations in management, recreation, and wildlife will find tough competition for jobs.

It is more difficult to estimate the possibilities for graduates of forest science programs. These students usually intend to seek advanced degrees and then teach or do research. Opportunities in these areas fluctuate wildly and are more dependent on the individual than on the state of the economy.

Veteran foresters often suggest that newly graduated foresters work in the field for a while before choosing a specialization for their master's degree. They also encourage undergraduates to load up on courses that will make them attractive to employers: engineering, wood science, communications, business, and international economics. The three Rs (reading, writing, and arithmetic) are important. So are the "three Cs" of computing, calculating, and computer operation.

Forestry is different from many other professions in that techniques vary with location. A forester's job in the tall timber of the Northwest may be quite different from a forester's job in the pines of the South. Both management practices and markets vary with the region. The availability of jobs also varies with the region.

In the United States, the North Central region is often the best place for graduating seniors to look for a job, followed by the South, the West, and the Northeast. The availability of jobs is related to the health of the market. You should consider this location factor when choosing a forestry school. Just as job opportunities are specific to a region, forestry education also varies with location. Forestry schools use local woods and mills as laboratories and classrooms. A forester learning the trade in California will get a different education than someone studying in Georgia or New York. If at all possible, pick a school in an area where you would like to work, and try to choose an area with a healthy and growing forest products economy.

Jobs for Forest Technicians

Almost everything said here about job opportunities for foresters holds true for forest technicians and aides. These paraprofessionals usually are graduates of a one- or two-year forestry course. Since they assist professional foresters, jobs for technicians and foresters generally are most abundant in the same areas.

The Forest Service is the single largest employer of forestry technicians. Logging, lumber, and paper companies are other major employers.

The Department of Labor expects slow growth in employment of forest technicians in the near future. Private industry will continue to provide most of the jobs. Some observers expect technicians to continue taking over a large number of the forester's traditional duties. As with foresters, there often are more graduates than jobs, so there is competition for technician positions.

Jobs for Wood Technologists

Graduates of a wood technology program should continue to find jobs available in the coming years. Wood utilization science is even more diverse than the management and production segments of forestry. Some portion of the industry is always growing.

Wood technologists may work for lumber or paper companies. They may find jobs with manufacturers of veneer, particleboard, laminated beams, plywood, furniture, sporting goods, prefabricated homes, chemical by-products, or containers. The list is practically endless.

Wood scientists may go into teaching or research and development. They, like technologists, may work for private industry or the government. They may specialize in such fields as wood, chemistry, or wood engineering.

Since the uses of wood touch every part of our society, there will always be jobs for those who know how to make the most of this material. The outlook for graduates of wood technology curricula is optimistic.

Effects of the Economy on Forestry

The status of the forestry industry depends directly on the health of the general economy. In a recession, foresters are among the first to feel the pinch. Much of the forest products industry is dependent upon the housing market. The construction of new homes requires all kinds of boards, veneers, plywood, and other wood products. When the economy is ailing, the housing market suffers. This throws the whole forestry industry into a tailspin. Mills close. The work force is cut back. Research and development activities are curtailed.

But what of the future? Eventually, when the economy improves, the housing market will pick up again. All those people who could not buy a house during the recession might then be able to afford one. It may never be as great a market as it once was. Houses may be smaller, using less wood. With a healthier economy, the paper industry, which held up well throughout the recession, should be even stronger. The markets for all those thousands of products made with wood should grow. That would improve demand for raw materials (trees) and create new jobs for foresters.

In the long run, it is worth remembering that even if the economy is unpredictable, one thing is certain: as the population grows, there will be an increasing need for wood products. This need will be met only through intense management of our forest resources, and that requires foresters.

Some calculate that population and market demands in the next few years will call for duplication of everything that has ever been built in the United States. The demand for timber will increase. There will be similar escalations in other forest-related needs. Water requirements will rise, big-game hunting will increase, freshwater fishing and outdoor recreation needs will jump. All of these increased demands depend upon well-managed forests.

Jobs Are Where You Find Them

Many opportunities for foresters fall outside of the traditional government and forest products industry. Other kinds of businesses own or administer woodlands and need foresters to manage them. These businesses include utility companies, railroads, petroleum and mining companies, sport and civic clubs, investment groups, steel companies and other heavy industry, recreation and housing developments, and private schools.

Other businesses—such as banks, insurance companies, agribusiness firms, and highway construction firms—sometimes hire staff foresters to advise on forestry matters.

Earnings

Salaries are difficult to predict. The inflation rate and shifting market scene can cause salaries to change rapidly. What is more, the salary you earn depends upon the company, the area, your skills and experience, your specialization, and just plain luck.

According to the *Occupational Outlook Handbook*, published by the Bureau of Labor Statistics, in 2001, entry-level foresters with a bachelor's degree earned an average of $23,776 to $30,035 annually. Those with a master's degree entered the job market with salaries between $30,035 and $42,783, while foresters with a Ph.D. started at around $52,162 or even $61,451 if working in research positions. Those employed by government agencies tend to fall at the lower end of this range, while foresters employed with private industry often draw the higher salaries.

Salaries also tend to rise with additional years of experience or when cost-of-living raises are provided. Foresters who move into supervisory or management positions also tend to earn higher salaries. In 2001, the average salary for federal foresters was $55,006. Forest products technologists earned an average of $71,572. By contrast, the average salary for rangeland managers was $50,715.

According to the Bureau of Labor Statistics, annual salaries for beginning forestry technicians started at $17,483, $19,453, or $22,251, depending on education and experience. As with foresters, higher salaries generally are paid by private industry or in certain parts of the country where the prevailing local pay level may be higher.

If an accurate salary figure is important to you in planning your career, call the nearest Federal Job Information Center and ask about current salaries. Remember, too, that there are other ways to be rewarded for your work. Some companies offer low-cost housing and fringe benefits such as free insurance and educational assistance. Many foresters are provided with a car or truck to drive on the job. Uniforms or other work clothes sometimes are provided. There also is the satisfaction of doing a job that is important to society and the pleasure of working close to nature.

9

Breaking In

The time has come to look for that first job in forestry. As with most careers, there are certain skills that you must have. Your education has given you a good foundation for your new profession, and any experience you have acquired through summer jobs and work-study will stand you in good stead. If you are prepared, persistent, and creative in your approach to looking for employment, you have a much better chance of finding a satisfying first job. As one forest products company executive put it: "Jobs in this field aren't growing on trees. You can't just walk out and pick one. But there are jobs available to the graduate with grades, maturity, and hustle."

When top-level executives of the largest forest products companies were asked to name the traits they looked for in job applicants, the executives agreed that the following qualities were important:

- Well-roundedness
- A good academic and extracurricular record

- A progressive record of applicable work experience
- Leadership abilities
- The ability to work well with limited supervision
- Trustworthiness
- A desire to excel
- The ability to get along with people

These are all traits that must be developed early.

Prepare While in School

Your future employers will be very interested in your academic career. It will be their best source of information regarding how you have spent the last few years. What will potential employers be interested in?

- **Choice of major.** Depending on the nature of the company or government agency, some specializations will be more of a bonus than others will.

- **Grade point average.** In a competitive job market, academic history is often the determining factor. On the other hand, some employers are less interested in grades than in other factors.

- **Extracurricular involvement.** Employers assume that students who are genuinely interested in forestry will be involved in related activities and organizations. Forestry students can find social and educational activities in the forestry club, and the Society of American Foresters has student chapters on campus.

- **Leadership development.** The more involved you are in student organizations, the better. These provide you with opportunities to develop leadership skills and identity. Seek experience in public speaking, organizing meetings, and group dynamics.

- **Summer employment.** Whenever possible, find summer jobs in your field of study, even if they don't pay as well as flipping burgers! Both the experience and the contacts you make will be extremely valuable in the long run.

- **Independent studies.** Get out on your own to tour nearby forestry operations. Spend time in the woods observing and learning. Read the professional journals and other publications that deal with your area of interest. Attend short courses, conferences, and symposia.

- **Use of career counseling services.** Most colleges and universities have career and placement assistance departments. Attend the courses they offer on how to find a job. These departments will assist with résumé preparation, letters of introduction, and job interview techniques. Participate in as many on-campus interviews as you can. The experience will be invaluable, and you might get an employment offer.

- **Communication skills development.** Employers need good communicators. Work on developing communication skills throughout your college years. Are there writing or speech classes that you can sign up for? Seek out any and all activities where you can strengthen your communication skills.

Finding a Job in Forestry

If you have followed the guidelines above and have properly prepared yourself during school, you should be able to find that first job. However, the job-hunting process itself can be stressful and intimidating the first time around. Here are several tips to make job hunting easier and more successful:

- Know and be able to articulate your particular interests in the forestry field. You should have a career goal and some idea of which type of employer will offer the job experience and training you need to reach your goal.

- Start your job search early. Do not wait until you graduate. In fact, part of your self-study activities in your final year should include learning all you can about the employment market and how to find a job.

- Contact foresters whom you have worked for or met during your school years. Ask them about the employment situation. Make as many personal, face-to-face contacts as possible. You can do this by attending meetings or by making trips on your own to visit woodland managers, forest consultants, state foresters, Forest Service foresters, and others. Attend as many on-campus interview sessions as possible.

- Keep up-to-date on current events in forestry. Learn about current policy issues. Keep abreast of the national economy and federal and state budget issues. This will give you a clue as to where the jobs are or are not. Also read *Journal of Forestry* or other periodicals that have current information about jobs and the economy.

• Learn about professional associations and employment agencies that offer employment assistance in the field of forestry. The Society of American Foresters (SAF) operates an employment referral service and assists members who are job-hunting. Available positions are advertised in each monthly issue of the *Journal of Forestry*. SAF members also may place "position wanted" ads in the journal.

• Obtain the most recent directories of employers' addresses. When you contact an employer, try to reach the head forester. He or she may have the position of woodlands manager, chief forester, district forester, or a similar title. If you do not address a specific individual, your application may get lost in the shuffle.

• Prepare a concise and well-written résumé. Carry it at all times. If you are a recent graduate, a one-page résumé should be adequate. Your résumé should highlight your background and your objectives for the future. It should capture the interest of the reader. Consider having your résumé printed by a professional printer.

• Research the employer before you interview. Know what the company does and where it operates. Learn the names of key personnel.

• Maintain constant contact with your college or university placement office. And don't forget to watch your forestry school's job bulletin board, too.

• Stay active in the job search. Do not give up. Be assertive and aggressive.

Women in Forestry

The traditional image of forestry was of a rugged, he-man occupation unsuitable for women, but that image no longer fits. Modern professional forestry is a broad field demanding a variety of skills and interests. And women are entering the field in increasing numbers. In recent years, 20 percent of undergraduates in forestry were women; in graduate school, women made up 50 percent of students.

Are the new women foresters encountering any resistance from men in the field? One woman, a professional forester who has dealt with a variety of employers, says that some adjustments are still being made: "Because of the profession's traditional masculine image, women did not enter forestry as early or as easily as they did other professions. However, the number of women in forestry is steadily growing, and it is my observation that they have gained considerable acceptance in the past ten years. I still observe that women feel more pressure to prove themselves, and that men may at first feel uncomfortable working with women. This strikes me as normal human behavior. Both men and women in forestry are still adjusting to each other."

Many of the women entering forestry are choosing specialties like communications, interpretation, wildlife management, and outdoor recreation. These specialties require less fieldwork than disciplines like engineering and timber management, and they may also be less bound by tradition. International forestry is another growing field for women, especially when working with development projects that target women in other countries.

Women may choose options with less fieldwork but should not consider the physical part of forestry an obstacle. One forester says:

"Fieldwork remains an important aspect of many foresters' routines, but few tasks actually require individual muscle. Good physical condition is more valuable than brute strength. For example, it's generally more important to be able to walk long distances in rough terrain than to swing an ax all day. But women shouldn't necessarily rule themselves out from the few jobs that do involve really tough physical work. I had a girlfriend who was the foreman of a logging crew and frequently ran a chainsaw all day long."

The proper image of the modern forester is simply a professional doing an important job. That image fits any qualified person, male or female.

Applying for Jobs with the U.S. Federal Government

All permanent positions with the Forest Service, agencies of the Department of the Interior, and other U.S. federal government organizations are Civil Service jobs. Becoming eligible for a permanent Civil Service position can be a long and tedious exercise. If you hope to work for the government, start this process early.

The first challenge is getting your name on the proper lists. The Office of Personnel Management keeps a file of all job openings and lists of eligible candidates. This agency was created under the Civil Service Reform Act of 1978. It manages the entire federal workforce. The U.S. Civil Service Commission, which formerly had this responsibility, was abolished.

These eligibility lists are made based on competitive examinations that rate experience and education. Each applicant is given a numerical rating. Those with the highest numbers are the first to be picked for available jobs.

More information can be obtained from the nearest Federal Job Information Center. The information center staff can tell you which job lists are open, when examinations are scheduled, qualifications required for certain jobs, salary levels, and other important information. These centers are located in major metropolitan areas across the country. They are listed under "U.S. Government" in the white pages of local phone directories. In addition, federal job openings are posted in state job service offices.

APPENDIX A

Forestry Organizations

THERE ARE SEVERAL major forestry organizations that offer a wealth of information and assistance to those interested in the field.

American Forests

The American Forests is the oldest and largest forestry organization in the country. Originally called the American Forestry Association, it was founded in 1875 by a group of citizens concerned about the uncontrolled exploitation of American forests. Since then the organization has worked with millions of citizens, hundreds of members of Congress, and numerous notables in the field of conservation and forestry. American Forests now has about forty thousand members.

American Forests takes pride in its long-standing role as a leader in the conservation education field. It began calling for conservation of our natural resources before most people had even heard of the term *conservation*. It was an early advocate of the protection and management of the federal forest reserves.

American Forests helped inspire Congress to create the Forest Service. Beginning in 1900, the association began a campaign to establish forest reserves in the eastern United States. This effort eventually was rewarded by the Weeks Act of 1911, which initiated action to set aside forest reserves in the eastern states. Until then, all of the forest designated as public had been in the West. The Weeks Act also initiated a national forestry policy and cooperation among states on forestry matters. For the next seven decades, American Forests was at the forefront of every major effort to pass natural resource legislation.

As other organizations began taking up the cause of conservation, American Forests often found itself in the role of arbitrator or referee. American Forests promotes balance between resource use and preservation. It also advocates planned multiple use of natural resources. As the citizen's voice for sound resource management, American Forests is well known in Congress. The association's headquarters are in Washington, D.C., allowing access to Capitol Hill. Executive staff members of American Forests often testify before Congress on resource legislation.

The members of American Forests have one thing in common—an interest in America's forestlands. Some, of course, are professional foresters. Many are involved in the wood products industry. The majority have nonforestry careers but are interested in natural resource management. Members receive the influential monthly magazine *American Forests*. This journal covers a wide range of resource issues. It has been reporting on the nation's resource policies for almost a century.

American Forests is involved in a number of cooperative projects with industry, government, and citizens groups. These projects, such as Global ReLeaf, all are designed to increase public awareness of natural resource problems and to encourage public participation in the solutions.

The positions that American Forests has taken cover the entire catalog of natural resources: timber, minerals, water, soils, fish and wildlife, grazing land, and wilderness. In each area, American Forests is working to ensure that the resource is used responsibly. The goal is balance between meeting today's resource needs and conserving resources for the future.

For more information, write to this address:

American Forests
P.O. Box 2000
Washington, DC 20013
americanforests.org

The headquarters of American Forests is located at:

910 Seventeenth St. NW, Ste. 600
Washington, DC 20006

Society of American Foresters

The Society of American Foresters represents the profession of forestry in the United States. This organization was created in 1900 by a group of farsighted pioneer foresters. Gifford Pinchot was the first president. Pinchot, often considered the "father of American forestry," was one of the very first professional foresters in the country. He was also the first dean at the Yale forestry school.

The SAF has seventeen thousand members. Most foresters who are actively working in the profession and interested in keeping abreast of new developments belong to SAF. Members include industry and government foresters, researchers, administrators, educators, consultants, forest technicians, and forestry students. To be eligible for full membership, one must be a graduate of an accredited forestry school with a bachelor's or higher degree or have at least three years of experience practicing in forestry or in a closely related field. There are special membership categories for forestry students and forestry technicians.

SAF is nonprofit and nonpolitical. It strives to advance the science, technology, education, and practices of forestry. Members are expected to obey a strict code of professional conduct. Through the years, these principles have brought the forestry profession the respect it deserves.

The society also is a unifying body. It allows foresters to speak to the public, or to legislators, with one voice. It unites forestry students with working professionals. As a nonpartisan organization, SAF helps bring together those with differing viewpoints on forest management. At local or national SAF meetings, foresters have a chance to meet others in the field and exchange ideas.

One of the society's most important functions is evaluating forestry programs at universities. SAF monitors the evolving forestry scene and suggests curriculum changes to schools of forestry. A committee of working foresters and educators establishes a tough set of standards. Those schools that can meet the standards are accredited by the society. By attending an accredited school, students can be assured that they are getting the education they need. By hiring a graduate of an accredited school, employers know they are getting an individual with a balanced education.

The society has thirty-four regional sections in the United States with 210 chapters. SAF has many Canadian members who belong to the nearest U.S. chapter. Forestry school students would do well to get involved with the SAF as early as possible.

The SAF headquarters is in Bethesda, Maryland, near Washington, D.C., which allows SAF staff close contact with federal agencies and other resource conservation organizations.

The society publishes the *Journal of Forestry*, a monthly magazine covering all branches of professional forestry. This excellent journal has the largest circulation of any magazine of its type in the world. Foresters rely on SAF's quarterly journals of applied forestry to keep them abreast of new developments. SAF also produces a quarterly magazine, *Forest Science*, which is devoted to research in forestry and related fields.

For further information, write to this address:

Society of American Foresters
5400 Grosvenor La.
Bethesda, MD 20814-2198
safnet.org

American Forest and Paper Association

The American Forest and Paper Association is the national trade association of the forest, pulp, paper, paperboard, and wood products industry. It was founded in 1993, merging the National Forest Products Association and the American Paper Institute. Its members collectively manufacture over 80 percent of the paper, wood, and forest products produced in the United States. A full-time staff administers the association's programs from its Washington, D.C., office.

The association develops long-range information and education programs to promote public awareness of the importance of forest resources. It provides expert advice to landowners interested in growing trees and sponsors a nationwide "tree farm" system. It runs general advertisements in the media, produces films, writes information brochures, and sponsors Project Learning Tree, a respected environmental education project. Its regular publication is *American Tree Farmer.*

The association's primary goal is to facilitate the exchange of information among the wood products industries, tree farmers, foresters, and the public. For more information, write to this address:

American Forest and Paper Association
1111 Nineteenth St. NW, Ste. 800
Washington, DC 20036
afandpa.org

Resource Conservation Organizations

There are many citizen organizations involved in forest conservation. Most of them have information and education designed to involve the public in natural-resource-use policy decisions. There also are organizations representing professional resource managers. These two types of national organizations include the following groups, each of which is followed by its website:

- The Conservation Foundation, theconservationfoundation.org
- Friends of the Earth, foe.org
- The Izaak Walton League of America, Inc., iwla.org
- National Arbor Day Foundation, arborday.org
- National Association of State Foresters, stateforesters.org

- National Wildlife Federation, nwf.org
- Natural Resources Council of America, naturalresourcescouncil.org
- Natural Resources Defense Council, nrdc.org
- The Nature Conservancy, http://nature.org
- Sierra Club, sierraclub.org
- Society for Range Management, rangelands.org
- Soil and Water Conservation Society, swcs.org
- Western Forestry and Conservation Association, westernforestry.org
- The Wilderness Society, wilderness.org
- Wildlife Management Institute, wildlifemanagementinstitute.org
- The Wildlife Society, wildlife.org
- World Resources Institute, wri.org/wri
- Worldwatch Institute, worldwatch.org

Trade Organizations

There are dozens of forest products industry coalitions and trade associations. Many of these have staffs and conduct information, legislative, education, and other programs. Some are organized just to unite the member companies. Others have forestry programs that involve the public. A partial list of trade organizations follows:

United States

Northwest Forestry Association (Forest Industries) (NFA)
1500 SW First Ave., Ste. 700
Portland, OR 97201
woodcom.com/woodcom/nfa

Southern Forest Products Association (Forest Industries) (SFPA)
P.O. Box 641700
Kenner, LA 70064
sfpa.org

Western Wood Products Association (Forest Industries) (WWPA)
522 SW Fifth Ave., Ste. 500
Portland, OR 97204
wwpa.org

Canada

Association of British Columbia Professional Foresters
1030-1188 W. Georgia St.
Vancouver, BC V6E 4A2
rpf-bc.org

Canadian Forestry Association
185 Somerset St. W
Ottawa, ON K2P OJ2
canadianforestry.com/eng

Canadian Wood Council
99 Bank St., Ste. 400
Ottawa, ON K1P 6B9
cwc.ca

Other Useful Addresses

Associated Landscape Contractors of America
150 Elden St., Ste. 270
Herndon, VA 20170
alca.org

Forest Products Society
2801 Marshall Ct.
Madison, WI 53705
forestprod.org

National Association of State Foresters
Hall of the States
444 N. Capitol St., Ste. 540
Washington, DC 20001
stateforesters.org

Society for Range Management
445 Union Blvd., Ste. 230
Lakewood, CO 80228
rangelands.org

USDA Forest Service
P.O. Box 96090
Washington, DC 20090-6090
fs.fed.us

U.S. Department of the Interior
1849 C St. NW
Washington, DC 20240
doiu.nbc.gov
The Department of the Interior has links to all the agencies within the DOI, including the Bureau of Indian Affairs, Bureau of Land Management, National Park Service, and U.S. Fish and Wildlife Service.

U.S. Office of Personnel Management
1900 E St. NW
Washington, DC 20415-0001
opm.gov

Appendix B
Forest Service Addresses

Addresses for the Forest Service headquarters as well as its field offices and research stations are provided below.

Headquarters

Forest Service
U.S. Department of Agriculture
Sidney R. Yates Federal Bldg.
201 Fourteenth St. at Independence Ave. SW
Washington, DC 20250

Field Offices

USDA Forest Service
Northern Region (R-1)
Federal Bldg.
200 Broadway
P.O. Box 7669
Missoula, MT 59807-7669

USDA Forest Service
Rocky Mountain Region (R-2)
740 Simms St.
Golden, CO 80401
P.O. Box 25127
Lakewood, CO 80225

USDA Forest Service
Southwestern Region (R-3)
Federal Bldg.
517 Gold Ave. SW
Albuquerque, NM 87102

USDA Forest Service
Intermountain Region (R-4)
Federal Bldg.
324 Twenty-Fifth St.
Ogden, UT 84401-2310

USDA Forest Service
Pacific Southwest Region (R-5)
1323 Club Dr.
Vallejo, CA 94592

USDA Forest Service
Pacific Northwest Region (R-6)
333 SW First Ave.
P.O. Box 3623
Portland, OR 97208

USDA Forest Service
Southern Region (R-8)
1720 Peachtree Rd. NW
Atlanta, GA 30367

USDA Forest Service
Eastern Region (R-9)
310 W. Wisconsin Ave., Rm. 500
Milwaukee, WI 53203

USDA Forest Service
Alaska Region (R-10)
709 W. Ninth St.
P.O. Box 21628
Juneau, AK 99802-1628

USDA Forest Service
Northeastern Area—S&PF
11 Campus Dr.
Newtown Square, PA 19073

Research Stations

USDA Forest Service
North Central Research Station
1992 Folwell Ave.
St. Paul, MN 55108

USDA Forest Service
Northeastern Forest Research Station
11 Campus Dr.
Newtown Square, PA 19073

USDA Forest Service
Pacific Northwest Research Station
333 SW First Ave.
P.O. Box 3890
Portland, OR 97208

USDA Forest Service
Pacific Southwest Research Station
800 Buchanan St., West Bldg.
Albany, CA 94710-0011
and
P.O. Box 245
Berkeley, CA 94701-0245

USDA Forest Service
Rocky Mountain Research Station
240 W. Prospect Rd.
Fort Collins, CO 80526-2098

USDA Forest Service
Southern Research Station
200 Weaver Blvd.
P.O. Box 2680
Asheville, NC 28802

USDA Forest Service
Forest Products Laboratory
One Gifford Pinchot Dr.
Madison, WI 53705-2398

USDA Forest Service
International Institute of Tropical Forestry
Call Box 25000
Rio Piedras, PR 00928-5000

UPR Experimental Station Grounds
Botanical Garden
Rio Piedras, PR 00928

APPENDIX C

Forest Technician Schools in the United States and Canada

THE FOLLOWING SCHOOLS offer educational programs leading to a two-year associate degree in forest technology or the equivalent. The programs here are those that are recognized by the Society of American Foresters.

Other schools not listed here may also offer similar programs.

United States

Schools in the United States include the following:

Alabama

Lurleen B. Wallace State Junior College
Forest Technology Program
Andalusia, AL 36420
http://lbw.edu

California

Reedley College
Landscape, Agriculture, and Natural Resources Department
Reedley, CA 93654
rc.cc.ca.us

Florida

Lake City Community College
Forest Management Department
Lake City, FL 32025-8703
lakecity.cc.fl.us

Georgia

Abraham Baldwin Agriculture College
Division of Agriculture and Forest Resources
Tifton, GA 31794-2693
http://stallion.abac.peachnet.edu

Illinois

Southeastern Illinois College
Forest Technology Department
Harrisburg, IL 62946
sic.cc.il.us

Maine

University of Maine–Fort Kent
Forest Technology Program
Fort Kent, ME 04743
umfk.maine.edu

Maryland

Allegany College
Forest Technology Program
Cumberland, MD 21502
ac.cc.md.us

Michigan

Michigan Technological University
School of Technology
Houghton, MI 49931
mtu.edu

Minnesota

Vermilion Community College
Forest Technology Program
Ely, MN 55731
vcc.mnscu.edu

New Hampshire

University of New Hampshire
Thompson School
Durham, NH 03824
unh.edu

New York

Paul Smith's College of Arts and Sciences
Forestry Division
Paul Smiths, NY 12970
paulsmiths.edu

SUNY College of Environmental Science and Forestry
Forest Technology Program
Wanakena, NY 13695
esf.edu

North Carolina

Haywood Community College
Division of Natural Resources
Clyde, NC 28721
haywood.cc.nc.us

Ohio

Hocking College
Forest Technology
Nelsonville, OH 45764
hocking.edu

Oklahoma

Eastern Oklahoma State College
Forestry Department
Wilburton, OK 74578
eosc.cc.ok.us

Oregon

Central Oregon Community College
Forest Technology Program
Bend, OR 97701
cocc.edu

Pennsylvania

Pennsylvania College of Technology
Forest Technology Department
Williamsport, PA 17701
pct.edu

Pennsylvania State University–Mont Alto
Forest Technology Program
Mont Alto, PA 17237
ma.psu.edu

South Carolina

Horry-Georgetown Technical College
Forestry Department
Georgetown, SC 29440
hor.tec.sc.us

Virginia

Dabney S. Lancaster Community College
Forest Technology Department
Clifton Forge, VA 24422
dl.cc.va.us

Washington

Green River Community College
Technology Division
Auburn, WA 98092
greenriver.ctc.edu

Spokane Community College
Natural Resources Department
Spokane, WA 99207
scc.spokane.cc.wa.us

West Virginia

Glenville State College
Department of Forest Technology
Glenville, WV 26351
glenville.edu

Canada

Schools in Canada include the following:

New Brunswick

Maritime Forest Ranger School
Fredericton, NB
Canada E3B 6H6
mfrs.nb.ca

Newfoundland

College of the North Atlantic
Forest Resources Technology Program
Corner Brook, NF
Canada A2H 6H6
northatlantic.nf.ca

Northwest Territories

Arctic College
Natural Resources Technology Program
Fort Smith, NT
Canada X0E 0P0
nac.nu.ca

Appendix D

Universities with Forestry Programs

Listed below are schools that offer a complete degree program in forestry. Some also offer graduate degree programs. U.S. programs are accredited by the Society of American Foresters. There are other accrediting agencies in various states and in other areas; additional schools and colleges continue to develop programs. Students interested in more up-to-date accreditation information should check with local, state, and provincial school authorities and with the Society of American Foresters.

United States

Schools in the United States include the following:

Alabama

Alabama A&M University
Center for Forestry and Ecology
Normal, AL 35762
aamu.edu

Auburn University
School of Forestry and Wildlife Sciences
Auburn, AL 36849-5418
auburn.edu

Alaska

University of Alaska
Department of Forest Sciences
Fairbanks, AK 99755
uaf.edu

Arizona

Northern Arizona University
School of Forestry
Flagstaff, AZ 86011
nau.edu

Arkansas

University of Arkansas at Monticello
School of Forest Resources
Monticello, AR 71655
uamont.edu

California

California Polytechnic State University
Natural Resources Management Department
San Luis Obispo, CA 93407
calpoly.edu

Humboldt State University
Department of Forestry and Watershed Management
Arcata, CA 95521
humboldt.edu

University of California-Berkeley
College of Natural Resources
Berkeley, CA 94720
berkeley.edu

Colorado

Colorado State University
Department of Forest Sciences
Fort Collins, CO 80525-1470
colostate.edu

Connecticut

Yale University
School of Forestry and Environmental Studies
New Haven, CT 06511
yale.edu

Florida

University of Florida
School of Forest Resources and Conservation
Gainesville, FL 32611-0410
ufl.edu

Georgia

University of Georgia
Warnell School of Forest Resources
Athens, GA 30602-2152
uga.edu

Idaho

University of Idaho
Department of Forest Resources
Moscow, ID 83843-1133
uidaho.edu

Illinois

Southern Illinois University
Department of Forestry
Carbondale, IL 62901-4411
siu.edu

University of Illinois
Department of Natural Resources and Environmental Sciences
Urbana, IL 61801
uiuc.edu

Indiana

Purdue University
Department of Forestry and Natural Resources
West Lafayette, IN 47907-1159
purdue.edu

Iowa

Iowa State University
Department of Natural Resource Ecology and Management
Ames, IA 50011
iastate.edu

Kentucky

University of Kentucky
Department of Forestry
Lexington, KY 40546-0073
uky.edu

Louisiana

Louisiana State University
School of Renewable Natural Resources
Baton Rouge, LA 70803-6200
lsu.edu

Louisiana Tech University
School of Forestry
Ruston, LA 71272
latech.edu

Maine

University of Maine
College of Natural Sciences, Forestry, and Agriculture
Orono, ME 04469
umaine.edu

Massachusetts

University of Massachusetts
Department of Natural Resources Conservation
Amherst, MA 01003-4210
umass.edu

Michigan

Michigan State University
Department of Forestry
East Lansing, MI 48824-1222
msu.edu

Michigan Technological University
School of Forestry and Wood Products
Houghton, MI 49931
mtu.edu

University of Michigan
School of Natural Resources and Environment
Ann Arbor, MI 48109-1115
umich.edu

Minnesota

University of Minnesota
College of Natural Resources
St. Paul, MN 55108-1030
umn.edu

Mississippi

Mississippi State University
College of Forest Resources
Mississippi State, MI 39762-9680
msstate.edu

Missouri

University of Missouri
School of Natural Resources
Columbia, MO 65211
missouri.edu

Montana

University of Montana
School of Forestry
Missoula, MT 59812
umt.edu

New Hampshire

University of New Hampshire
Department of Natural Resources
Durham, NH 03824-3589
unh.edu

New York

SUNY College of Environmental Science and Forestry
Faculty of Forestry and Natural Resources Management
Syracuse, NY 13210-2788
esf.edu

North Carolina

Duke University
Nicholas School of the Environment
Durham, NC 27708-0329
duke.edu

North Carolina State University
College of Forest Resources
Raleigh, NC 27695-8008
ncsu.edu

Ohio

The Ohio State University
School of Natural Resources
Columbus, OH 43210-1085
acs.ohio-state.edu

Oklahoma

Oklahoma State University
Department of Forestry
Stillwater, OK 74078
okstate.edu

Oregon

Oregon State University
College of Forestry
Corvallis, OR 97331-5704
orst.edu

Pennsylvania

The Pennsylvania State University
School of Forest Resources
University Park, PA 16802-4300
psu.edu

South Carolina

Clemson University
Department of Forest Resources
Clemson, SC 29634-0306
clemson.edu

Tennessee

University of Tennessee
Department of Forestry, Wildlife, and Fisheries
Knoxville, TN 37901-1071
utk.edu

Texas

Stephen F. Austin State University
College of Forestry
Nacogdoches, TX 75962-6109
sfasu.edu

Texas A&M University
Department of Forest Science
College Station, TX 77843-2135
tamu.edu

Utah

Utah State University
Department of Forest Resources
Logan, UT 84322-5215
usu.edu

Vermont

University of Vermont
School of Natural Resources
Burlington, VT 05405-0088
uvm.edu

Virginia

Virginia Polytechnic Institute and State University
Department of Forestry
College of Natural Resources
Blacksburg, VA 24061
vt.edu

Washington

University of Washington
College of Forest Resources
Seattle, WA 98195
washington.edu

Washington State University
Department of Natural Resource Sciences
Pullman, WA 99164-6410
wsu.edu

West Virginia

West Virginia University
Division of Forestry
Morgantown, WV 26506
wvu.edu

Wisconsin

University of Wisconsin–Madison
Department of Forest Ecology and Management
Madison, WI 53706-1598
wisc.edu

University of Wisconsin–Stevens Point
College of Natural Resources
Stevens Point, WI 54481
uwsp.edu

Canada

Schools in Canada include the following:

Alberta

University of Alberta
114 St.-89 Ave.
Edmonton, AL
Canada T6G 2M7
ualberta.ca

British Columbia

British Columbia Institute of Technology
3700 Willingdon Ave.
Burnaby, BC
Canada V5G 3H2
bcit.ca

University of British Columbia
2329 W. Mall
Vancouver, BC
Canada V6T 1Z4
ubc.ca

New Brunswick

University of New Brunswick
100 Tucker Park Rd.
P.O. Box 5050
St. John, NB
Canada E2L 4L5
unb.ca

Ontario

Lakehead University
955 Oliver Rd.
Thunder Bay, ON
Canada P7B 5E1
lakeheadu.ca

University of Toronto
Toronto, ON
Canada M5S 1A1
utoronto.ca

Appendix E

Addresses of State Foresters

National Association of State Foresters
444 N. Capitol St., Ste. 540
Washington, DC 20001
stateforesters.org

Alabama

Alabama Forestry Commission
513 Madison Ave.
Montgomery, AL 36130

Alaska

Division of Forestry
State Forester's Office
550 W. Seventh Ave., Ste. 1450
Anchorage, AK 99501

American Samoa

Forestry Program Manager
P.O. Box 5319
ASCC/AHNR
Pago Pago, American Samoa 96799

Arizona

State Land Department
2901 W. Pinnacle Rd.
Phoenix 85027-1002

Arkansas

Arkansas Forestry Commission
3821 W. Roosevelt Rd.
Little Rock, AR 72204-6396

California

Department of Forestry and Fire Protection
P.O. Box 944246
1416 Ninth St., Rm. 1505
Sacramento, CA 94244-2460

Colorado

Colorado State Forest Service
Colorado State University
203 Forestry Bldg.
Fort Collins, CO 80523

Connecticut

Division of Forestry
79 Elm St.
Hartford, CT 06106

Delaware

Delaware Forest Service
2320 S. DuPont Highway
Dover, DE 19901

District of Columbia

Urban Forestry Administration
District Department of Transportation
1105 O St. SE
Washington, DC 20003

Florida

Division of Forestry
3125 Conner Blvd.
Tallahassee, FL 32399-1650

Georgia

Georgia Forestry Commission
P.O. Box 819
Macon, GA 31202-0819

Guam

Forestry and Soil Resources Division
192 Dairy Rd.
Mangilao, Guam 96923

Hawaii

Division of Forestry and Wildlife
1151 Punchbowl St.
Honolulu, HI 96813

Idaho

Idaho Department of Lands
954 W. Jefferson St.
Boise, ID 83720-0050

Illinois

Division of Forest Resources
2005 Round Barn Rd.
Champaign, IL 61821

Indiana

Department of Natural Resources
402 W. Washington St., Rm. W296
Indianapolis, IN 46204

Iowa

Department of Natural Resources
Wallace Office Building
E. Ninth and Grand Ave.
Des Moines, IA 50319

Kansas

Kansas Forest Service
2610 Claflin Rd.
Manhattan, KS 66502-2798

Kentucky

Kentucky Division of Forestry
627 Comanche Trail
Frankfort, KY 40601

Louisiana

Office of Forestry
P.O. Box 1628
Baton Rouge, LA 70821

Maine

Maine Forest Service
22 State House Station
Harlow Bldg.
Augusta, ME 04333

Marshall Islands

Ministry of Resources and Development
Coconut St.
P.O. Box 1727
Majuro, Republic of the Marshall Islands 96960

Maryland

Forest Service
Department of Natural Resources—Forest Service
580 Taylor Ave., E-1
Annapolis, MD 21401

Massachusetts

Division of Forest and Parks
P.O. Box 1433
Pittsfield, MA 01202

Michigan

Michigan Department of Natural Resources
Mason Bldg., 8th Fl.
Box 30452
Lansing, MI 48909-7952

Micronesia

Department of Economic Affairs
P.O. Box PS-12
Palikir, Pohnpei
Federated States of Micronesia 96941

Minnesota

Division of Forestry
500 Lafayette Rd.
St. Paul, MN 55155-4044

Mississippi

Mississippi Forestry Commission
301 N. Lamar St., Ste. 300
Jackson, MS 39201

Missouri

Missouri Department of Conservation
P.O. Box 180
Jefferson City, MO 65102

Montana

Department of Natural Resources and Conservation
Forestry Division
2705 Spurgin Rd.
Missoula, MT 59801

Nebraska

Nebraska Forest Service
Plant Industries Building, Rm. 101
Lincoln, NE 68583-0815

Nevada

Division of Forestry
1201 Johnson St., Ste. D
Carson City, NV 89706-3048

New Hampshire

Division of Forests and Lands
Box 1856
172 Pembroke Rd.
Concord, NH 03302-1856

New Jersey

State Forestry Service
P.O. 404
Trenton, NJ 08625-0404

New Mexico

Forestry and Resources Conservation Division
P.O. Box 1948
Santa Fe, NM 87504-1948

New York

New York State Department of Environment Conservation
625 Broadway
Albany, NY 12233-4250

North Carolina

North Carolina Division of Forest Resources
1616 Mail Service Center
Raleigh, NC 27699

North Dakota

North Dakota Forest Service
307 First St.
Bottineau, ND 58318-1100

Northern Mariana Islands

Territorial Forester
Department of Lands and Natural Resources
CNMI, P.O. Box 10007
Siapan, Northern Mariana Islands 96950

Ohio

Division of Forestry
1855 Fountain Square Ct., H-1
Columbus, OH 43224

Oklahoma

Oklahoma Department of Agriculture
Forestry Services
P.O. Box 528804
Oklahoma City, OK 73152-3864

Oregon

Oregon Department of Forestry
2600 State St.
Salem, OR 97310

Palau

Palau Agriculture and Forestry
Erenguul St.
P.O. Box 460
Koror, Palau 96940

Pennsylvania

Bureau of Forestry
P.O. Box 8552
Harrisburg, PA 17105-8552

Puerto Rico

Forest Service Bureau—DNER
P.O. Box 9066600
Puerta de Tierra
San Juan, PR 00906-6600

Rhode Island

Division of Forest Environment
1037 Hartford Pike
North Scituate, RI 02857

South Carolina

South Carolina Forestry Commission
P.O. Box 21707
Columbia, SC 29221

South Dakota

Resource Conservation and Forestry
Foss Bldg.
523 E. Capitol Ave.
Pierre, SD 57501

Tennessee

Tennessee Department of Agriculture
Division of Forestry
P.O. Box 40627
Melrose Station
Nashville, TN 37204

Texas

Texas Forest Service
301 Tarnow Dr., Ste. 364
College Station, TX 77840-7896

Utah

Department of Natural Resources
1594 W. North Temple, Ste. 3520
Salt Lake City, UT 84114-5703

Vermont

Department of Forest, Parks, and Recreation
103 S. Main St.
Waterbury, VT 05671-0601

Virgin Islands

Department of Agriculture
Estate Lower Love
Kings Hill
St. Croix, VI 00850

Virginia

Virginia Department of Forestry
900 Natural Resources Dr., Ste. 800
Charlottesville, VA 22903

Washington

Department of Natural Resources
Box 47001
1111 Washington St.
Olympia, WA 98504-7001

West Virginia

West Virginia Division of Forestry
1900 Kanawha Blvd., E
Charleston, WV 25305-0180

Wisconsin

Department of Natural Resources, Division of Forestry
P.O. Box 7921
Madison, WI 53707

Wyoming

Wyoming State Forestry Division
1100 W. Twenty-Second St.
Cheyenne, WY 82002

About the Author

Christopher M. Wille has written about the environment for more than twenty years. He received his bachelor of science degree in wildlife management from Oregon State University and his master's degree in English/creative writing from Central State University in Edmond, Oklahoma.

From 1983 to 1989, Mr. Wille was a vice president for the National Audubon Society. He presently directs the Tropical Newsbureau, a project of the Rainforest Alliance, and lives in San Jose, Costa Rica.